AF271621

The Second Quantum Revolution

Volume 3

Foundational Transformation of Quantum Electrodynamics.

Two Branches of Electromagnetism

Victor Vaguine, Ph.D.

ConsReality® Press

ISBN: 978-1-936795-21-5
Library of Congress Control Number: 2020947319

Published in the United States of America
ConsReality® Press.
Dallas, Texas
www.consreality.com

Publisher's Cataloging-in-Publication data

Victor Vaguine. Ph.D.
The Second Quantum Revolution.
Volume 3
Foundational transformation of quantum electrodynamics.
Two branches of electromagnetism / Vaguine, Victor

ISBN: 978-1-936795-21-5
Includes bibliographical references and Index

1. Quantum theory. 2. Physics. 3. Philosophy 4. Experiments. I. Title

QC679.6.V34 2020
539.1 - dc23   2020947319

Editing by Galena Miller-Horii.
Cover illustration Copyright   2020 by ConsReality, Inc.
Cover design by Andre Vaguine/ConsReality, Inc.
Illustrations by Andre Vaguine/ConsReality, Inc.
Book design and production by Andre Vaguine/ConsReality, Inc.

Edition 1.0

ConsReality is a trademark of ConsReality, Inc.

Dedication

To Vera.

Acknowledgements

My sincere and profound appreciation to my son Andre Vaguine for his dedication and unwavering support.

CONTENTS

List of Postulates for Volume Three
(continued)

Postulate 45. Elastic Interaction and Explanation of "Charge Screening".

Postulate 46. The definition of the physical c-ring.

Postulate 47. The General Compton Conditions and their application to the intrinsic electron (prior the corset action).

Postulate 48. Spin of the intrinsic electron.

Postulate 49. Gyromagnetic ratio of intrinsic electron in both, classical and quantum, presentations.

Postulate 50. The c-ring model for the fermions.

Postulate 51. The Position Parameter of the intrinsic electron.

Postulate 52. The constants l_c and L_a are fundamental constants of the intrinsic electron.

Prologue to Volume Three

*Science is the uncompromising search
for the truth in objective reality.*

*No force in the world can stop fun-
damental scientific truth.*

Aphysical Quantum Mechanics is the result of thirteen years of my scientific work as an independent scientist.

Aphysical Quantum Mechanics is a deeper and more profound quantum theory and the origin of the Second Quantum Revolution. Aphysical Quantum Mechanics is published in three separate volumes under the title The Second Quantum Revolution [1], [2], [3]:

- Volume One: "Foundational Transformation of Quantum Mechanics. Aphysical Quantum Mechanics as Deeper Quantum Theory. Elementary Consciousness of Elementary Particles".

- Volume Two: "Foundational Transformation of Quantum Optics. AQM Theory of the Photon".

- Volume Three: "Foundational Transformation of Quantum Electrodynamics. Two Branches of Electromagnetism".

In principle, one would not expect AQM to produce immediate scientific revolutions in other branches of fundamental physics, such as quantum optics and elementary particle physics, but that is exactly what has happened. Both of these branches—quantum optics and especially elementary particle physics—have undergone the dramatic foundational transformation.

The AQM is based on avalanche of fundamental concepts in the form of 30 postulates in Volume One, 14 postulates in Volume Two and 32 postulates in Volume Three. Those fundamental concepts are synergistic and reinforce each other.

The first quantum revolution has never been completed. For all those decades since 1927, QM has remained an incomplete theory with huge fundamental gaps and fundamental pieces missing.

My scientific work sets in motion the Second Quantum Revolution, the genuine and profound one. It brings at the outset an avalanche of fundamental concepts and scientific fundamental discoveries.

AQM expands fundamental understanding of the quantum reality by postulating two more fundamental categories in addition to the physical category: the aphysical category and the elementary consciousness of elementary particles. There is no way one can comprehend the quantum reality without these two additional fundamental categories.

I have discovered and explained three-dimensional inner structures and self-masses for each class of elementary particles including new classes. Every step of their individual interactions in spacetime dynamics can be visualized in the human mind. New principles and new particles (in fact, new classes of elementary particles!) are postulated, with a plethora of new fundamental properties.

It has turned out that "the bread and butter particles", such as the electron, the photon, the neutrino, the proton, and the neutron, have been left behind and, surprisingly, hardly explored fundamentally. Whatever we knew about these particles prior to the Second Quantum Revolution was only the tip of the iceberg.

Volume Three is devoted to the electron and to the neutrino. Therefore, it is appropriate to bring upfront the direct comparison of the QM electron and the QM neutrino with the AQM electron and the AQM neutrino as the confirmation of the above "bread and butter" statement. It is presented as Addendum to Prologue for Volume Three.

The paradigmatic power of Aphysical Quantum Mechanics is such that I was able to uncover 53 fundamental misconceptions/absurdities in the Standard Model theory of particle physics; and to make 27 fundamental scientific discoveries presented in all three Volumes.

AQM opens the floodgates of new physics. It brings democracy into foundational physics. Graduates and undergraduates will be able to select new physics problems to work on among the thousands immediately available. Hundreds of doctoral dissertations will be written based on AQM. And most importantly, for a change, the general public will be brought along.

In his book, "The Structure of Scientific Revolutions", [4] Thomas S. Kuhn observes that genuine scientific revolutions are manifested by the destructive-constructive paradigm, meaning the massive destruction of old information and its replacement by new.

Not all pre-revolution theories and concepts are destroyed. Those that are along the natural scientific trajectory toward a deeper understanding of objective reality survive and flourish.

That is exactly what happening in the Second Quantum Revolution driven by Aphysical Quantum Mechanics. It is not by chance

15

the Second Quantum Revolution is originated outside the scientific establishment.

The Copenhagen interpretation is replaced by the AQM paradigm. All known longstanding quantum enigmas, paradoxes, and mysteries are resolved. As far as QM is concerned, its role is about to shift from foundational physics to a useful mathematical quantum application tool for the calculation of statistical values, with some paradigmatic limitations recognized and applied.

Here are some principal points of Aphysical Quantum Mechanics:

- Quantum reality consists of three fundamental categories: physical, aphysical, and elementary consciousness.

- One of the core principles of AQM is the spacetime visualization of individual elementary particles and individual elementary quantum processes. *Physics without spacetime visualization has no future.*

- All elementary particles have non-zero size. All elementary particles have the three-dimensional inner structures of perfect geometry consisting of *the physical energy c-ring(s), the aphysical energy cylinder(s), and elementary consciousness.*

- The three-dimensional inner structure of perfect geometry is in conflict with the Heisenberg probabilistic uncertainty principle, the principle not relevant to individual quantum entities and individual quantum processes.

- The inner structure of the fundamental elementary particle has a single physical c-ring and a single aphysical cylinder; the inner structure of the composite particle has two or more physical c-rings and two or more aphysical cylinders.

- Spin is the expression of elementary consciousness. Spin is the actual rotation in three-dimensional space at Compton angular velocity and peripheral velocity v = c. Spin is the mode of existence of the elementary particles and is eternal over their lifetimes. There are no elementary particles without spin, obvious or hidden. Spin can be visualized in a classical way. The elementary consciousness is the force driving spin. *No elementary consciousness – no spin – no self-mass.*

- Elementary particles are particles, never waves. Wave property is a probabilistic expression only.

- Free elementary particles travel in space along their individual trajectories, absolutely defined. The full-fledged elementary particle has a single trajectory. The self-entangled elementary particle has several trajectories but only one carries the physical substance. All other trajectories are aphysical.

- The elementary particle has a defined position in its frame of self-reference with zero momentum. This is also valid for bosons in their frames of self-reference v = c. The origin of self-mass ("self-energy") for fermions and bosons (yes, for bosons!) is established and explained. Both fermions and bosons have self-mass. There is no more division of "massless" and "massive". Yes, bosons are no longer "massless". Bosons have self-mass and zero kinetic energy in their frames of self-reference, v = c. Virtual energy does not exist.

- A self-entangled elementary particle can be in several places at the same time, physically always in one place and aphysically in all the others. Just prior to physical-physical interaction, a particle is instantaneously reconstructed to the full-fledged quantum state regardless of distances separating the host and a-fractions, with no violation of special relativity.

- After physical-physical interaction of one elementary particle with another elementary particle or with a macroscopic entity, such as a measurement device, no entanglement between them remains in the aftermath of their interaction. Without much ado, the perennial "measurement problem" exists no longer. There is an exception under some special conditions such as BE condensation at microKelvin temperature and superconductivity.

- AQM has expanded understanding of determinism and causality in the description of the quantum world. Yes, indeed, there are "hidden variables", such as position parameters. For example, no two muons are identical. The difference in their lifetimes is explained by the difference in initial values of their position parameters acquired at their formation. *A decay of the individual particle is the deterministic process.*

- AQM explains that, indeed, there are fundamental sources of quantum irreducible randomness such as (1) in the selection of the host in self-entanglement; and (2) in the position of the physical c-ring relative to the aphysical cylinder in the inner structure of the elementary particle.

- In the world of elementary particles, there is harmonious coexistence of the classicality and the quantum. The classical-quantum divide is a myth. *In AQM, the classicality has found its perfect expression.*

- Quantum laws are not universal. The quantum reality is a subset of the objective reality. Gravity is not part of the quantum reality. Humans and cats are not a subset of quantum reality.

- In my firm view, the consciousness is the most important and enormously significant fundamental category. However, there is very little understanding of the consciousness as a fundamental category of objective reality. Consciousness cannot be derived from anything and neither can it be reduced to anything.

- Each elementary particle has non-zero size.

A *pre-corset* c-ring Compton radius of elementary particles can be calculated in a classical way. In case of the electron, prior its formation, a pre-corset c-ring classical Compton radius r_c is equal to $2 \times 3.86 \times 10^{-13}$ meters. After the corset action, the c-ring is "instantaneously" and "dramatically" reduced to extremely small *quantum Compton radius* r_q, estimated in the range of $10^{-22} - 10^{-23}$ meters.

- Each elementary particle has spin rotating with the speed of light at the Compton angular velocity. The speed of light has a special significance in the world of elementary particles. It is incorporated into the inner structure of each elementary particle.

- SM declares that the electron is the basic fermion of electromagnetism. This is an SM fundamental misconception, among many others.

- According to AQM, the electron is a composite particle consisting of two constituents: the intrinsic electron and the electron neutrino ("duo-electrino"). Electron magnetic moment is the result of electric charge rotation and *weak electric charge rotation.* There is no "anomaly" in the electron magnetic moment. The so-called "anomaly" is actually the electron neutrino magnetic moment. Calculation of "anomaly" on a basis of QED contributions is scientifically "illegitimate".

- For over a hundred years, scientists have been searching for the electron physical model, with no success. I have discovered the correct electron physical model, which in itself is a historical milestone. The discovery explains a plethora of new physical-aphysical properties of the electron including the origin of self-mass, spin, the electron formation, electrostatic and magnetostatic field configurations, combined magnetic moment, the electric and weak charge fractionation, and three-dimensional configuration of the Cooper electron pairs in superconductivity.

- It is SM fundamental misconception that all fermions have spin ½. The misconception comes from the accepted notion that there is only one Planck constant, it is one of a kind, and is applicable to all fundamental forces. More careful analysis shows that the origin of Planck constant is electromagnetism. I postulated that each fundamental force has its own Planck constant.

- According to AQM, the intrinsic neutrino $\hat{v}$ and the intrinsic antineutrino $\hat{\bar{v}}$ are the fundamental fermions of weak electromagnetism which is a second branch of electromagnetism with spin ½ $\hbar_w$ where $\hbar_w$ is Planck constant of weak electromagnetism. The intrinsic neutrino $\hat{v}$ carries exclusively negative weak electric charge $-w$ and the intrinsic antineutrino $\hat{\bar{v}}$ carries exclusively positive weak charge $+w$. A value of weak electric charge w is established with great precision.

- On the other hand, the commonly accepted neutrino in particle physics is in fact the duo-fermion of weak electromagnetism with two constituents: the intrinsic neutrino ($\hat{v}$) and the *reversed* (meaning, opposite helicity) intrinsic antineutrino ($\hat{\bar{v}}^R$), presented as

$$v = \left\{ \begin{array}{c} \hat{v} \\ \hat{\bar{v}}^R \end{array} \right\},$$

and the antineutrino, presented as

$$\bar{v} = \left\{ \begin{array}{c} \hat{v}^R \\ \hat{\bar{v}} \end{array} \right\}.$$

- The neutrino has two opposing spins, $+½\ \hbar_w$ *and* $-½\ \hbar_w$, resulting in spin S = 0. *The neutrino has zero spin.* It is a hidden spin.

- I postulated that the neutrino is a single field fermion with magnetostatic energy only (see chapter 4). The neutrino is a stable particle, as stable as the electron. The neutrino can travel long distances, as has been confirmed by observations from super-novae and the Sun.

- Contrary to SM, the neutrino is not a lepton.

- The fundamental role of the neutrino in the world of elementary particles has been discovered and explained. The neutrino provides the corset function for relative or absolute stability of leptons.

- According to SM, electron neutrino mass is zero, although nowadays few physicists believe that. For many decades there has been ongoing theoretical and experimental effort toward resolving the issue of electron neutrino mass, with marginal results so far. The present consensus is that the electron neutrino mass is less than or equal to 0.120 eV. This is another SM misconception.

The longstanding problem with electron neutrino mass issue is solved in AQM with the precision of five digits. AQM provides a correct value to the electron neutrino self-mass. *The electron neutrino self-mass is equal to 296.118 828 eV.*

This is one of the greatest AQM triumphs.

A new challenge for the mathematically inclined physicists with intuition is to develop a mathematical formalism for the description of individual elementary quantum processes in spacetime dynamics. This is what is on the horizon and coming to quantum physics in the near future. However, even without new mathematics, I was able to reconstruct in detail the spacetime dynamics of selected individual elementary quantum interactions and processes, such as the muon decay and the subsequent formation of the electron.

My style, somewhat repetitive and redundant is not by chance.

Fundamental concepts are evolving from volume to volume, reaching maturity and perfection at the end.

The presented material in Volumes One, Two and Three is original, has never been published, and requires few references.

♦ ♦ ♦

Postscript: By publishing The Second Quantum Revolution I have fulfilled my promise given in my book Prologue to Super Quantum Mechanics (2012) [5].

♦ ♦ ♦

Addendum for Prologue

Electron and Neutrino: Definition and Properties, SM vs AQM.

Standard Model (SM) versus Aphysical Quantum Mechanics (AQM)

1   Standard Model

1.1   The Electron according to the Standard Model

The electron is the basic fermion of electromagnetism. The electron is point-like and structureless. It has mass of 0.511 MeV/c^2 of unknown nature, negative electric charge (–e), spin $S = \frac{1}{2}\,\hbar$, and anomalous magnetic moment M equal to one Bohr magneton × $(1+\alpha)$, where

$\alpha = 0.001\ 159\ 652\ 188$

The origin of electron self-mass is either unknown or is attempted to be explained by the Higgs mechanism.

The Pauli exclusion principle is applied to electrons – two electrons cannot occupy the same quantum state.

The positron is the antiparticle of the electron.

The SM electron is simplistic in spite of over 100 years of research and experimentation. Most of electron properties are overlooked.

Simplistic understanding of the electron as a fundamental fermion of electromagnetism is one of the most damaging mental block in particle physics.

1.2 The Neutrino according to the Standard Model

A neutrino is a fermion that interacts only with the weak force and gravity.

Mass: < 0.120 eV, 95% confidence level, sum of 3 flavors.

Spin: ½ $\hbar$

Electric charge: zero e

Antiparticle: opposite helicity from particle

2 Aphysical Quantum Mechanics

2.1 The AQM electron is a composite fermion of the electromagnetism and the weak electromagnetism

The electron has non-zero size with classical Compton radius of $2 \times 3.86 \times 10^{-13}$ meters just prior to the electron formation, or quantum Compton radius with upper limit of 10^{-22} meters after the electron is formed. The electron has a three-dimensional inner structure consisting of two constituents: the intrinsic electron ($\hat{e}^-$) and the electron neutrino (ν_e) (see Figure P-1).

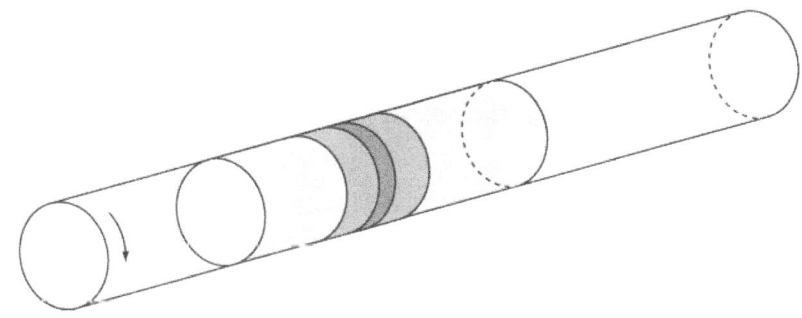

Figure P-1

The inner structure of the electron.

2.2 The intrinsic electron is the first electron constituent

The intrinsic electron ($\hat{e}^-$) is a fundamental fermion of the electromagnetism with spin S = ½ $\hbar$, negative electric charge (-e), self-mass of 0.511 MeV/c² and magnetic moment equal to one Bohr magneton.

The intrinsic electron inner structure consists of the physical c-ring with length l_c, and the aphysical cylinder with length L_a.

The intrinsic electron has electrostatic and magnetostatic fields described by classical Maxwell electromagnetism. Electrostatic energy is equal to magnetostatic energy.

2.3 The electron neutrino is the second electron constituent

The electron neutrino is a composite fermion of *weak electromagnetism*, consisting of the intrinsic neutrino ($\hat{v}$) with negative weak electric charge (-w), the intrinsic antineutrino ($\hat{\bar{v}}^R$) with positive weak electric charge (+w) of opposite helicity (a notation R means reversed). The intrinsic neutrino and the intrinsic antineutrino are fundamental fermions of the weak electromagnetism, exclusive carriers of weak electric charge. In AQM terminology the electron neutrino is the duo-electrino. It has duo configuration of its inner structure. The electron neutrino is a Majorana particle.

The electron neutrino (v_e) is a single field elementary particle. It has weak magnetostatic field and zero weak electrostatic field.

Neutrino field configuration is described by Maxwell weak electromagnetism.

The electron neutrino has zero spin $S = 0$ (it has a hidden spin $S = \frac{1}{2} \hbar_w - \frac{1}{2} \hbar_w$, where $\hbar_w$ is Planck constant for weak force, $\hbar_w = \beta \hbar$, and $\beta = \frac{1}{2} \alpha$, and $\alpha = 0.001\ 159\ 652\ 188$).

The electron neutrino has weak electric charge zero, where one of its constituents, the intrinsic electron neutrino, has negative weak electric charge $(-\beta e)$, and another constituent, the intrinsic electron antineutrino, has positive weak electric charge $(+\beta e)$.

The electron neutrino has magnetic moment M equal to two weak Bohr magnetons (weak Bohr magneton = β Bohr magneton).

The electron neutrino inner structure consists of the physical c-ring with length l_c^v and the aphysical cylinder with length L_a^v.

The electron neutrino performs the corset function at the formation of the electron bringing the electron size to the level of 10^{-22} meters.

2.4 Conclusion

Electron spin $S = -\frac{1}{2} \hbar$.

The electron has negative electric charge $(-e)$.

Electron magnetic moment is a sum of intrinsic electron magnetic moment (one Bohr magneton) and electron neutrino magnetic moment (two weak Bohr magnetons). The electron has no anomaly in its magnetic moment. The problem of the anomalous electron magnetic moment is finally solved.

Electron self-mass is the sum of intrinsic electron self-mass of 0.511 MeV/c^2 and electron neutrino self-mass of 296 eV/c^2. In fact, on the basis of AQM, one can derive the electron neutrino self-mass with precision to six digits using experimental data of Dehmelt group (1987).

Electron neutrino self-energy is equal to 296.118826(4) eV. It is one of the greatest AQM triumphs.

The electron inner structure contains three position parameters:

- PP1 defines the location of the intrinsic electron c-ring relative to the intrinsic aphysical cylinder;

- PP2 defines the location of the neutrino c-ring relative to the electron c-ring; and

- PP3 defines the location of the neutrino c-ring relative to the neutrino aphysical cylinder.

The origin of the electron self-mass is the sum of electromagnetism and weak electromagnetism. Self-mass of the intrinsic electron can be computed on the basis of its inner structure geometry, electric charge density distribution, boundary conditions, and electrostatic and magnetostatic field distributions.

References for Prologue

[1]   Victor Vaguine, *The Second Quantum Revolution. Volume 1. Foundational Transformation of Quantum Mechanics. Aphysical Quantum Mechanics as Deeper Quantum Theory. Elementary Consciousness of Elementary Particles.* ConsReality Press, 2020.

[2]   Victor Vaguine, *The Second Quantum Revolution. Volume 2. Foundational Transformation of Quantum Optics. AQM Theory of the Photon.* ConsReality Press, 2020.

[3]   Victor Vaguine, *The Second Quantum Revolution. Volume 3. Foundational Transformation of Quantum Electrodynamics. Two Branches of Electromagnetism.* ConsReality Press, 2020.

[4]   Thomas S. Kuhn, *The Structure of Scientific Revolutions,* third edition, The University of Chicago Press, 1966.

[5]   Victor Vaguine, *Prologue to Super Quantum Mechanics.* ConsReality Press, 2012.

Chapter 1
History of the Electron and Background Information

"You know, it would be sufficient to really understand the electron" – Albert Einstein.

"Thus, the electron may have size and structure!" – Hans Dehmelt, 1989 Nobel Laureate.

"We will be considered the generation that left behind unsolved such essential problems as the electron self-energy" – Wolfgang Pauli, 1945 Nobel Laureate.

1.1 General Comments on the Electron

One would think that not much is left to discover about the electron. What a misconception! Aphysical Quantum Mechanics (AQM) expands dramatically the fundamental understanding of the electron including its three-dimensional composite inner structure, physical and aphysical properties, elementary consciousness, and the explanation of all electron related Quantum Mechanics (QM) enigmas. The electron is no longer enigmatic. It can be visualized in all details. Visualization is the strength of AQM. The challenge for mathematically inclined physicists with intuition is to develop a new mathematical formalism on the basis of AQM for the description of an individual elementary interaction of an individual electron with another individual elementary particle in spacetime dynamics. It would keep mathematically intuitive physicists productive for many decades. It is challenging, rewarding and exciting task, comprising for example, of detailed spacetime dynamics, the electron inner structure transformation, process of annihilation and reconstruction.

The Standard Model (SM) claims that the electron is the fundamental fermion of the electromagnetism. This is another SM misconception.

According to AQM, the electron is a composition fermion of the electromagnetism consisting of the intrinsic electron $\hat{e}^-$, the fundamental fermion of electromagnetism, and the electron neutrino ν_e of duo configuration.

1.2   Historical Models of the Electron

Electron was discovered in 1897 by J.J. Thomson. Today electron is one of the most studied elementary particles with its many properties discovered and experimentally measured, such as mass, electric charge, spin, magnetic moment, anomaly in magnetic moment, stability, and quantum properties.

Most of historical models of the electron are based on classical electrodynamics. From the time of its discovery, there has been ongoing effort to explain the origin of electron self-mass in terms of electromagnetism. All proposed classical models have failed.

For further discussion, I select only those classical electrodynamics models which are educational and a step in the right direction.

Here are three examples of classical electrodynamics models: the charged spinning sphere, the Goudsmit and Uhlenbeck concept, and the spinning uniformly charged ring.

1.3 The Charged Spinning Sphere as Electron Model

The spherical spinning electron model with electric charge density distributed uniformly on the rigid sphere surface was proposed by Abraham in 1902 (see Figure 1). At the relativistic limit, maximum angular velocity for the spinning sphere is equal to Compton angular velocity,

$$\omega = c/R$$

where R is radius of the sphere and c is the speed of light.

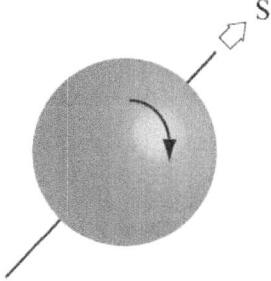

Figure 1

The charged spinning sphere as electron model

That means that the maximum linear velocity on the sphere equator is equal to the speed of light. With such maximum spinning, the performance of the spherical model is somewhat disappointment. It is unstable. The sphere would explode in the direction of the poles from electrostatic repulsive force with no opposing force. Spin and magnetic moment are below experimental values. The electromagnetic self-mass is only 75% of the experimental electron self-mass.

However, the spherical spinning model is not a complete failure. It brings some useful information and is a step in the right direction.

Here is what we have learned from the model:

- Electromagnetic energy can produce electron self-mass (self-energy), although not to the full extent.

- Spin and magnetic moments are below their experimental values.

- Stability is achieved only at the equator where the Compton angular velocity $\omega = c/R$ is applied. Here the repulsive electrostatic force is exactly balanced by the inward magnetostatic pinch force.

- Spin constitutes spinning in the classical way and can be visualized.

- If Nature ever decides to create such a creature as the spherical spinning electron, it would explode instantaneously in the direction of the poles, where the electrostatic repulsive force is totally unopposed.

1.4   Goudsmit and Uhlenbeck's Idea

In 1925 two Dutch graduate students, Samuel Goudsmit and George Uhlenbeck, attempted to explain electron magnetic moment. They put together the following basic thought: electron spin, electric charge, and magnetic moment are interrelated. They assumed that electron spin is not just a quantum parameter, but an actual spinning and rotating electric charge, thus producing electron magnetic moment. Thus, Goudsmit and Uhlenbeck postulated that the electron has the intrinsic classical spin and the related classical magnetic moment.

That was a step in the right direction. Rather than to capitalize on the idea and develop it further, the concept was met with skepticism from many eminent physicists including Pauli.

"But despite these quite reasonable objections, their model stubbornly continued to agree with experimental results!" – stated Frank Wilczek, a Nobel Prize winner [1], (see page 163).

1.5   The Charged Spinning Ring as Electron Model

Now, let us consider the spinning ring electron model proposed by David L. Bergman and J. Paul Wesley [2]. The proposed charged spinning ring is another electron model based on classical electrodynamics (see Figure 2). This model was totally ignored by the physics community. The spinning ring electron model has electric charge density uniformly distributed over the entire surface. The ring is spinning with Compton angular velocity

$$\omega = c/R \ ,$$

where R is Compton radius.

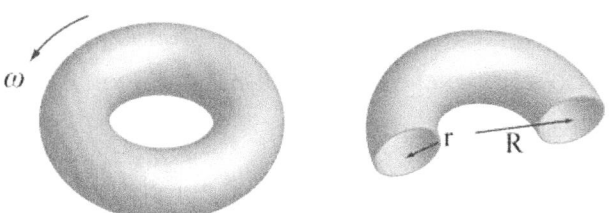

Figure 2

The charged spinning ring as electron model

The spinning ring has four parameters: ring radius R, inner radius r, electric charge e, and spinning Compton angular velocity

$$\omega = c/R$$

The surface of the ring is formed by uniformly distributed elementary electric charge (-*e*). The surface charge elements travel with tangential linear velocity in the range

$$1 - r/R \leq v/c \leq 1 + r/R.$$

This is problematic. Classical electrodynamics does not allow physical substance to exceed the speed of light. *Strictly speaking, the model is not scientifically viable.*

The reduction in value of inner radius *r* would not cure the deficiency unless *r* = 0. In such a case, the model collapses into a singularity.

However, contrary to the model reality, the authors made *a forceful assumption* – all surface elements travel with the speed of light, regardless. Then things begin to fall into right places.

The outward repulsive electrostatic force is balanced exactly with inward magnetostatic pinch force over the entire ring surface, thus making the model stable. Electrostatic energy E_E is equal exactly to magnetostatic energy E_H.

Spin of the electron is equal to ½ $\hbar$, magnetic moment is equal to one Bohr magneton, and self-mass is equal to electromagnetic mass $(E_E + E_H)/c^2$.

However, this model has to be rejected. As mentioned above, some surface elements exceed the speed of light regardless of the authors' forceful assumption. In other surface areas where velocity is less than the speed of light, there is no balance of opposing forces. In addition, the model has an extra geometrical parameter, inner radius *r*. The parameter *r* is not fixed, it is variable. It allows one to produce a whole spectrum of electron models, which is an absurdity. In the extreme case of $r/R = 0$, it brings the model to a singularity. The model is "fancy". Nature demonstrates to us, over and over again, *its majestic simplicity and sophistication.*

However, we have to be open-minded. Although the model is not viable, it shows certain conditions and is in the right direction towards the correct electron model based on classical electrodynamics. A viable model must have linear velocity on the surface equal to the speed of light over the entire charge surface:

(1) to achieve stability by balancing electrostatic repulsive inward pressure P_E at the surface with inward magnetostatic pinch pressure P_H

$$P_E = -P_H;$$

(2) to achieve equality of electrostatic energy E_E with magnetostatic energy E_H

$$E_E = E_H; \text{ and}$$

(3) to obtain correct values of spin, magnetic moment, and self-mass of electromagnetic nature.

The search for a viable electron model must continue.

1.6   The Electron Quantum Model according to the Standard Model

There is one more electron model to consider. My work is devoted to quantum mechanics issues. Even if I would prefer to, I cannot afford to skip the SM electron quantum model, which is described in all particle physics literature and is reluctantly accepted by quantum physicists.

The SM electron quantum model is point-like with no structure. Electron mass m and electric charge e are placed into an infinitely small point in space, thus bringing mass density and electric charge density to infinite values.

In the vacuum, electron is surrounded by cloud of virtual particles, such as electron-positron pairs producing what is called *the vacuum polarization*, thus causing the shielding effect for electron charge and making the effective charge, looking from a distance, smaller than its "true" value that exceeds electric charge *e* (see Figure 3 (a)).

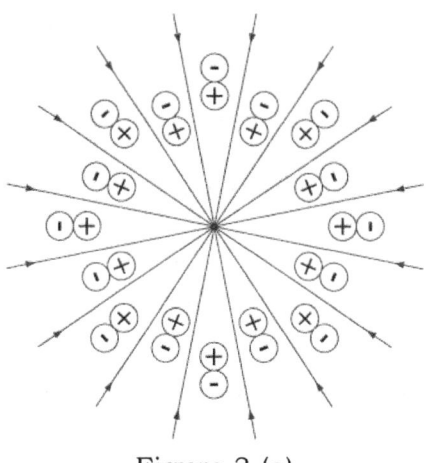

Figure 3 (a)

The SM quantum model of the electron

Other electron quantum parameters, such as spin and magnetic moment with empirically correct values, are attached to the point model by *proclamation*. There is no explanation for the origin of electron self-mass. The countless virtual electron-positron pairs, in principle, require their own vacuum polarization, thus bringing the model into infinite regress.

Obviously, this is not a scientific model. It is a nonsensical model. It is, as John Bell would have said, a model "for all practical purposes."[3]. It is a mathematical model, where one has to use renormalization technique by deducting one infinity from another infinity to obtain the result. The electron model is another SM misconception.

The issue of the vacuum polarization is suspect. It has to be totally re-examined. According to AQM, pairs of virtual particles arising spontaneously from the vacuum do not exist. Such mechanism violates the energy conservation law. Appealing to the Heisenberg uncertainty principle does not help. The uncertainty principle is a probabilistic principle. It applies to assemblies and not to individual quantum interactions.

At the first sign of difficulty, rather than compromise with half-baked ideas, it is much safer to stick to the integrity of the energy conservation law. *Science is uncompromising search for truth of objective reality.*

After all, haven't we learned something from history of physics? In 1930s, after the discovery of beta-decay, there was no explanation to beta-decay energy imbalance. Bohr proposed a compromise: the conservation energy law is not applicable to individual quantum interactions – it is valid only for assemblies. A similar situation exists with countless virtual particles arising from the vacuum. They violate the energy conservation law by borrowing energy from the vacuum for a tiny instant of time and then return the borrowed energy back. Such concept is entrenched in particle physics. It is part of quantum mindset. It is one of numerous fundamental misconceptions of the SM quantum model of the electron.

In elastic electron-electron interaction, involving, just as an example, only two colliding electrons, both electrons experience stress in each other's electrostatic fields. As a result, the electrons radiate photons (real photons) in the direction of their momenta causing repelling force additionally to Coulomb force, thus creating impression that electric charge is larger than it actually is (see Figure 3 (b)).

No vacuum polarization is needed for the explanation of "charge screening".

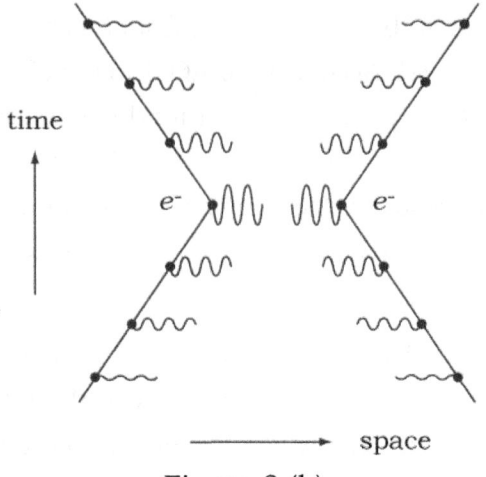

Figure 3 (b)

Electron-electron elastic interaction

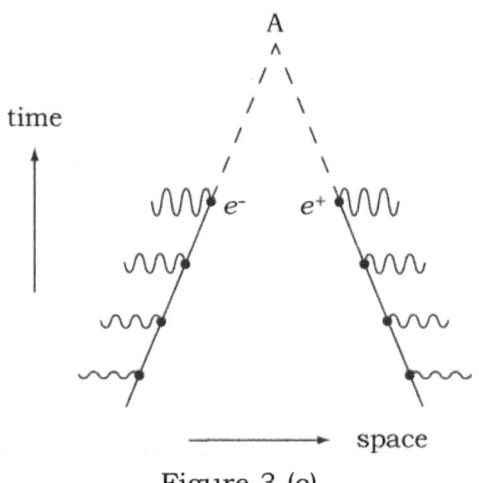

Figure 3 (c)

Electron-positron collision and their annihilation (point A)

In electron-positron collision prior their annihilation, both electron and positron radiate photons (real photons) in the direction opposite to their momenta, causing attractive force additional to Coulomb attractive force, thus creating impression that electric charge is larger than it actually is (see Figure 3 (c)).

Again, no vacuum polarization is needed for explanation of "charge screening".

Resulting kinetic energy in elastic collision is reduced by photon radiation, thus proving that the radiation is real and not virtual. There are no virtual photons.

♦ ♦ ♦

Postulate 45. *Elastic interaction and explanation of "charge screening".*

At the elastic electron-electron interaction electrons radiate photons (real photons) in the direction of their (electrons) momenta, thus adding additional repelling force to Coulomb force.

After such elastic interaction is completed the overall kinetic energy of electrons is reduced by amount of photon radiation energy.

At the electron-positron interaction prior their annihilation, electron and positron radiate photons (real photons) in the direction opposite of their (electron and positron) momenta, thus adding additional attractive force to Coulomb force.

One can call these photons as mediating photons. However, they are not intercepted by their partners.

That is how "charge screening" is explained. There is no need for vacuum polarization, which does not exist anyway.

♦ ♦ ♦

1.7   Toward a Viable Model of the Intrinsic Electron Based on Classical Electrodynamics

According to AQM, the electron has size and the inner structure.

In fact, according to AQM, *there are three types of electrons – the intrinsic electron, the duo-electron, and the electron.* The electron is a composite fermion of the electromagnetism. The intrinsic electron is the fundamental fermion of the electromagnetism and in its bound state is a constituent of the duo-electron and the electron.

Since 1902, when it was first proposed by Abraham, the spherical spinning model of the electron has been studied and rejected by many physicists. In 1904, Laurence proposed a revised model where the sphere was flattened along the direction of motion. It was also rejected. In 1905, Poincare proposed a non-electromagnetic force of unknown origin to balance electrostatic repulsive force. Eventually, the spherical spinning electron model was abandoned.

Let us re-examine the spinning spherical model in depth. At the equator, the forces are balanced between electrostatic repulsion and magnetostatic pinch. However, in the direction of poles, the electrostatic repulsive force is unopposed. If Nature were to create such an electron, it would instantaneously explode in the direction of poles (see Figure 4 (a)).

Rather than show impatience, which is amply demonstrated by great scientific minds, let us remove troublesome areas in the model, namely both semi-spheres, retain only the infinitely narrow equator strip, and then distribute uniformly the whole electric charge e along the equator (see Figure 4 (b)). As a result, we obtain a singularity model which is balanced and stable. This is a first extreme electrodynamic model of the intrinsic electron. After that, we proceed to the next step by stretching the singularity equator into a uniformly charged short section of the cylinder, the c-ring (see Figure 4 (c)). Voila! We have arrived at the electrodynamic model of the intrinsic electron where correct electromagnetic field configuration at electrodynamic parameters, including self-mass, spin, and magnetic moment, are achieved and can be calculated.

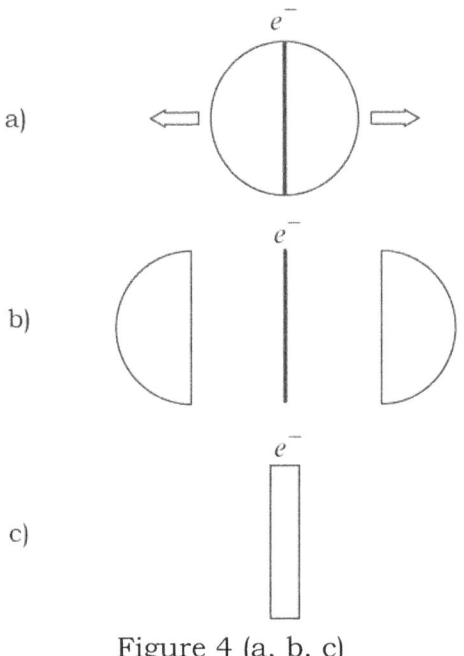

Figure 4 (a, b, c)

The transformation of the spherical model into the c-ring

The c-ring model is an amazing scientific discovery. Nature proves again its majestic simplicity and sophistication. This discovery could have been made before the Fifth Solvey Conference (1927). By then, Einstein' special relativity was established and well understood. *The discovery of the c-ring structure of the intrinsic electron would have changed the historical trajectory of quantum mechanics.*

Rather than using abstract mathematical tools, such as operators, propagators, commutation, wave function, observables, eigenvalues and eigenstates, which describe only statistical reality, quantum mechanics would be dealing with physical reality of the intrinsic electron in spacetime dynamics.

By now, the reader understands that the c-ring is only a part of the intrinsic electron inner structure. The remaining parts are the aphysical cylinder and the elementary consciousness. Furthermore, the complete electron is a composite elementary particle consisting of the intrinsic electron and the electron neutrino. In its term, the neutrino is the composite of the intrinsic neutrino and the intrinsic antineutrino in duo configuration.

For many decades, the electron has been left largely unexplored. A plethora of new electron properties is discovered by the author and included in Volume 3.

The question can be asked: "is the c-ring 100% electrodynamics and nothing else?" The answer is "all classical electrodynamics properties are included in the c-ring". The c-ring is more than just a classical electrodynamic design. There are other quantum properties such as self-entanglement and entanglement, and the ability of the c-ring to radiate photons when it is under stress, in situations such as the synchrotron radiation or elastic interaction with another charged particle.

1.8   A Simple Relativistic Test for Electrodynamic Electron Models

The AQM c-ring model is the only electrodynamic model which passes a simple relativistic test. All three electrodynamic electron models: the spinning sphere, the spinning ring, and the spinning c-ring with initial arbitrary spin orientation at a pre-relativistic velocity (v<<c), are subjected to acceleration to a velocity approaching the speed of light (c-v << c). Both, the sphere and the ring change their shapes, thus causing a non-uniform charge distribution over their surfaces, while the c-ring, after relativistic contraction, still remains in its c-ring form, thus preserving uniform charge distribution, although with a greater density (see Figure 5).

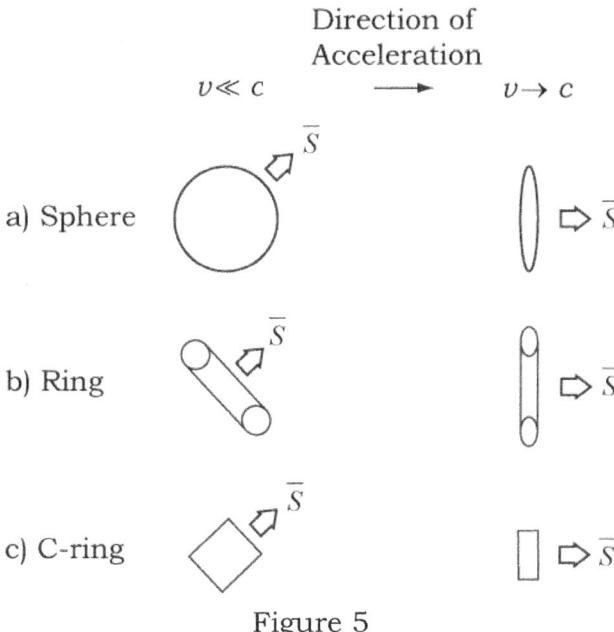

Figure 5
A simple relativistic test for electrodynamic electron models

During one of his visits to SLAC in 1970s, Richard Feynman was shown a high energy electron-electron colliding experiment. Feynman exclaimed enthusiastically that the collision of super-relativistic electrons looked like a collision of "pancakes." The reader, glancing at Figure 5, might notice that this statement implies the discredited spherical model of the electron. The correct statement is that electron-electron collision is the collision of electron c-rings, although most of colliding c-rings are only partially overlapped.

The radius of the electron c-ring does not experience relativistic contraction.

1.9   The AQM Intrinsic Electron Model
Survives the Charge Fractionation Test

The AQM intrinsic electron model is the only one among historical electrodynamics models that survives charge fractionation test: e, $\frac{2}{3} e$, $\frac{1}{3} e$. The intrinsic electron electric charge fractionation is shown in Figure 6.

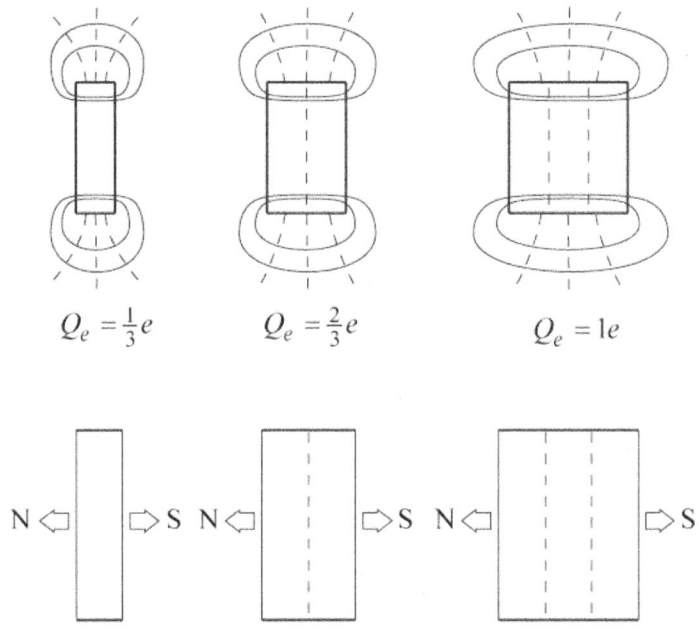

$$Q_e = \frac{1}{3} e \qquad\qquad Q_e = \frac{2}{3} e \qquad\qquad Q_e = 1e$$

Figure 6

A charge fractionation test for the AQM intrinsic electron model

The intrinsic electron c-ring electric charge fractionation is shown in Figure 7 (a): Similarly, fractionation of aphysical cylinder is shown in Figure 7 (b, c):

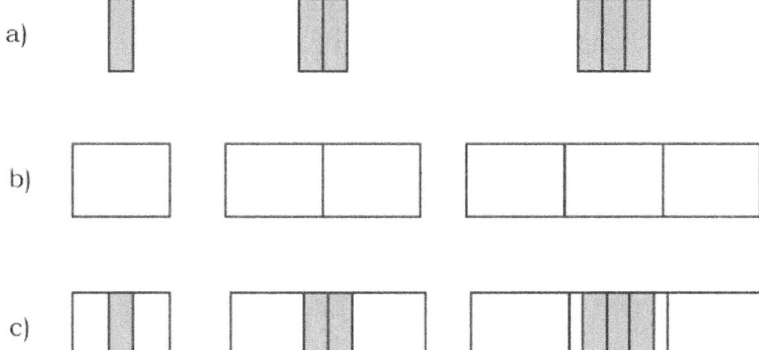

Figure 7 (a, b, c)

The AQM intrinsic electron fractionation

Here is the set of constants for the length of the c-ring,

$$l_c, \frac{2}{3} l_c, \frac{1}{3} l_c$$

and the set for the length of the aphysical cylinder,

$$L_a, \frac{2}{3} L_a, \frac{1}{3} L_a.$$

They are building blocks of leptons and quarks, as shown later. One can already see an initial glimpse into a revised system of elementary particles.

45

1.10 The AQM Electron C-ring Model Survives Cooper Pairs Test in Superconductivity

The existing theory of superconductivity is based on the formation of the electron pairs (Cooper pairs). An unanswered question remains – how can electrons attract each other overcoming the Coulomb repulsion at sufficiently low temperature when interactions of electrons with the vibrating crystal lattice is reduced?

AQM provides a straightforward explanation stating that there is no the Coulomb repulsion between an aligned pair of electrons as long as the vibrating energy of crystal lattice is sufficiently low, thus preventing disruption of the electron pair formation.

The process of formation occurs when two electron c-rings are aligned along their common axes with their magnetic moments in attraction mode – "north" meets "south" (see Figures 8 a, b).

In such configuration, the Coulomb repulsion between two electron c-rings rotating at equal Compton angular velocity and equal Compton radius does not exist.

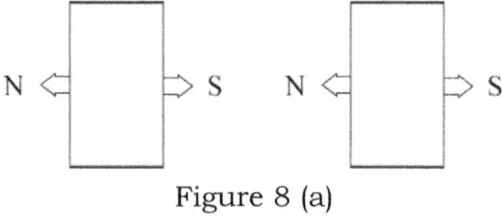

Figure 8 (a)

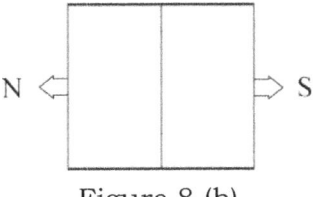

Figure 8 (b)

Ontological explanation of Cooper electron pairs

The pair formation is a statistical process proceeding fast at sufficiently low temperature when a binding energy between two magnetic moments is greater than a disrupting vibration energy of surrounding ions of the crystal lattice.

During electron pair formation, magnetic moments M1 and M2 and spins S1 and S2 are aligned, resulting in an electron pair with double spin S = 1 and double magnetic moment M equal two times of one Bohr magneton.

Such pair arrangement is energetically more advantageous as compared to a single electron. The formation process releases some energy, reduces Compton angular velocity, and increases Compton diameter. The released energy is transferred to the lattice and promptly removed from the system.

The produced pairs with spin S =1 acquire some bosonic properties. In fact, it is the assembly of entangled electron pairs with properties similar to the Bose-Einstein condensate (BE condensate). The assembly of Cooper pairs can be kept indefinitely while the BE condensate exists in micro-Kelvin temperature environment for just a few seconds. It is a matter of technology.

Aside from Cooper pairs, what other electron combinations are possible?

A combination with odd number of electrons is fermionic and cannot be formed according to the Pauli's exclusion principle. But the exclusion principle is a formal statement with no ontological explanation. "QM explains nothing."

The strength of AQM is in its ontology. It does not need the exclusion principle to explain why combinations with odd number of electrons, such as $2 + 1$, would not work. Compton diameters and angular Compton frequencies of the c-rings are slightly different between a pair and a single electron. For a formation to succeed, these properties must be absolutely identical. The only other viable combinations are multiple pairs of 2^n, where $n = 1, 2, 3$, etc.

As we keep reducing temperature beyond micro-Kelvin range toward the absolute zero, one should expect the electron formation of higher orders: 2, 2+2, 4+4, 8+8, and so on. The higher the order the more it is energetically advantageous.

Where is the limit? Eventually, at extremely low temperature, in the range of $10^{-12} - 10^{-18}$ K *all electrons of the assembly form the single structure of the highest order.* It is my educated guess.

Chapter 2
AQM Intrinsic Electron: the Inner Structure and Properties

2.1   The Inner Structure of the Intrinsic Electron

Pre-AQM physics does not tell us much about fundamentals of the electron. Surprisingly, from the time the electron was discovered in 1897, it was left mostly unexplored. In spite of extensive scientific effort over many decades to find the electron size, the inner structure, and the origin of self-mass, not much progress has been made. SM mistakenly presents the electron as a basic fermion of electromagnetism with no size, no inner structure, and no constituents. This is another example of SM fundamental misconception.

As shown in Chapter 8, according to AQM, the electron is a composite elementary particle, consisting of three fundamental constituents: the intrinsic electron $\hat{e}^-$, the intrinsic neutrino $\hat{\nu}$ and the intrinsic antineutrino $\hat{\bar{\nu}}$ with opposite helicity.

Chapter 8 of Volume 3 presents the complete AQM theory of the electron. Chapters 2, 3, and 5 of Volume 3 present description and detailed analysis of each individual constituent of the electron, such as the intrinsic electron, the intrinsic neutrino, and the neutrino.

AQM expands our fundamental understanding of the electron and brings forth a plethora of new properties.

The principal constituent of the electron is the intrinsic electron. The first step toward finding the correct intrinsic electron inner structure is to begin with the correct classical electrodynamic structure. Surprisingly, for many decades it has been staring in our face in a form of the c-ring. As shown in Figure 9, the intrinsic electron inner structure consists of the physical energy c-ring, the aphysical energy cylinder, and the elementary consciousness residing in the c-ring, where l_c is the length of the c-ring, L_a is the length of the aphysical cylinder, and PP is the position parameter. Both, l_c and L_a are constants of the same fundamental significance as Planck constant, electric charge, and the speed of light.

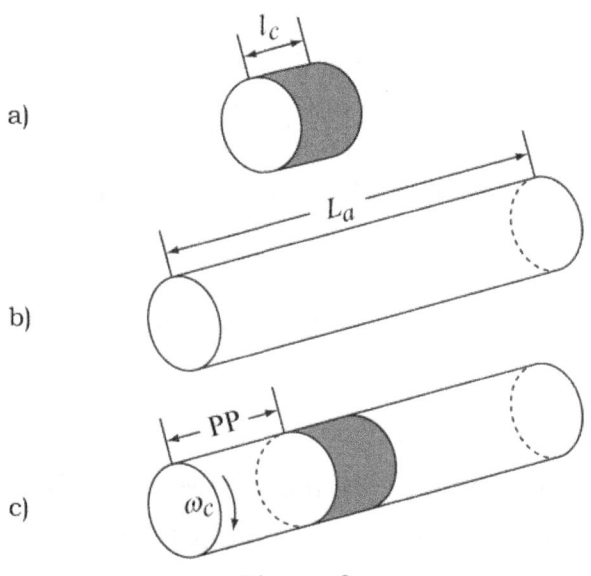

Figure 9
The inner structure of the intrinsic electron:
a) the physical c-ring
b) the aphysical cylinder (AC)
c) the inner structure

Postulate 46. *The definition of the physical c-ring:*

The physical c-ring is a section of the hollow cylinder with zero wall thickness.

In case of fermions, the c-ring has non-zero length. In case of bosons, the length of the c-ring is zero.

♦ ♦ ♦

I emphasize that there must be no confusion between "the ring" and "the c-ring". For comparison, the c-ring versus the ring are shown in Figure 10 (a,b). The ring (b) is not compatible with the intrinsic electron inner structure.

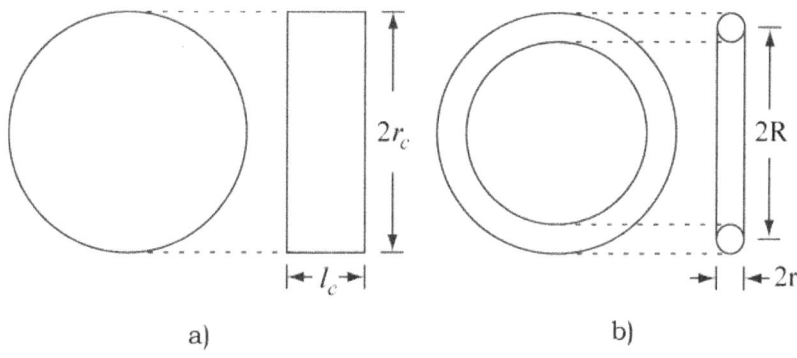

a)   b)

Figure 10 (a,b)

Comparison of the c-ring and the ring

c-ring. $\omega_c = c/r_c$.

Ring. $\omega = c/R$

The electrodynamic inner structure of the intrinsic electron is the physical c-ring. The c-ring consists of the cylindrical surface made from uniformly distributed electric charge e. The c-ring spins with the Compton angular velocity $\omega_c = c/r_c$ and peripheral linear velocity equal to the speed of light. In addition to classical electrodynamic properties, the c-ring also has some specific quantum properties, such as emission of real photons (called "virtual photons"), emission of synchrotron radiation photons, and radiation of electromagnetic energy.

2.2   General Compton Conditions as Basis for Design of the Intrinsic Electron

The Compton relation is well-known and defined at the relativistic limit of velocity $v = c$ as a relation between Compton angular velocity ω_c and Compton radius r_c:

$$\omega_c r_c = c$$

According to AQM, the Compton relation is as fundamental in elementary particle physics as Einstein's formula $m = E/c^2$. *The Compton relation is incorporated into the inner structure of every elementary particle.*

According to QM, electron spin should not be considered in a classical sense as something actually spinning – spin is a quantum parameter found experimentally and assigned *by proclamation.*

AQM asserts that the classical-quantum divide does not exist. It does not imply that quantum physics can be reduced to classical physics. In AQM, spin is considered as something actually spinning and can be visualized in human mind. The intrinsic electron spin is derived mathematically from the AQM model in a straightforward way.

The Compton relation is only a first step in the right direction.

AQM expands the Compton relation to a more comprehensive *General Compton Conditions*. The Compton relation is a subset of the General Compton Conditions. All elementary particles, fundamental and composite, are explained on the basis of the General Compton Conditions adapted to each particular class of elementary particles and to each particular fundamental force. There is no exception to this rule.

Using the intrinsic electron as an example, I define the fundamental meaning of the General Compton Conditions as applied specifically to the intrinsic electron c-ring in a form of postulate.

◆ ◆ ◆

Postulate 47. *The General Compton Conditions and their application to the intrinsic electron (prior the corset action):*

- *The intrinsic electron c-ring spins with a classical Compton angular velocity:*

$$\omega_c = c/r_c, \qquad (2\text{-}1)$$

  where r_c is a classical Compton radius.

- *The c-ring produces electrostatic field and generates magnetostatic field. Under the General Compton Conditions, electrostatic energy E_E is equal to magnetostatic energy E_H:*

$$E_E = E_H. \qquad (2\text{-}2)$$

- *Radial forces at the c-ring surface are balanced. Outward repulsive electrostatic pressure P_E is opposed equally by inward magnetostatic pinch pressure P_H over the entire c-ring surface:*

$$P_E = -P_H. \qquad (2\text{-}3)$$

- *There are no repulsive or attractive tangential forces along the c-ring surface. All elements of electric charge rotate in parallel to each other at the speed of light.*

- *The c-ring surface is made of negative electric charge (–e) with uniform charge density distribution and zero thickness.*

- *Electric charge is a special state of matter not yet recognized by science. By itself, it has no self-mass, gravitation, or inertia. It exists under the General Compton Conditions.*

- *Under stress, the intrinsic electron radiates photons with exponential intensity and exponential energy relative to applied stress.*

- *If accelerated, the intrinsic electron radiates electromagnetic energy.*

- *The AQM c-ring model of the intrinsic electron is mathematically accurate requiring no approximation. It is described by classical electrodynamics.*

◆ ◆ ◆

2.3   Electromagnetic Field Configuration of the Intrinsic Electron C-ring

The c-ring electromagnetic field configuration is described by classical electrodynamics. Boundary conditions are precisely defined. Negative electric charge is uniformly distributed on the surface of the c-ring producing electrostatic field. The c-ring spins with Compton angular velocity generating magnetostatic field. Electrostatic and magnetostatic field lines cross each other in space always perpendicular to each other, as shown in Figure 11.

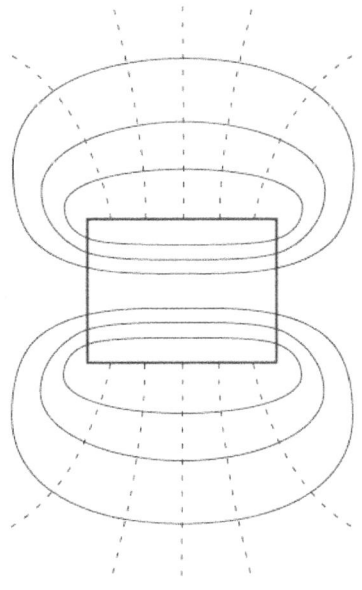

Figure 11

Electrostatic and magnetostatic field configurations for the intrinsic electron c-ring

Using computation, one can demonstrate that electrostatic and magnetostatic energies are equal: $E_E = E_H$. Immediately, a question arises "what length of the c-ring is to be assumed?"

With respect to the c-ring length l_c for the intrinsic electron, Nature assigns only one specific value. The c-ring length l_c is a fundamental constant. One cannot theoretically derive fundamental constants, such as the speed of light c in vacuum, Planck constant h, or elementary electric charge e. All of them are found experimentally. The c-ring length is also a fundamental constant. It can be derived on the basic of General Compton Conditions, Planck constant, and available experimental data, such as intrinsic electron self-mass.

The question of the c-ring length is even more profound. What are c-rings lengths for the intrinsic muon or the intrinsic tau? And what is c-ring length for the intrinsic electron with fractional electric charges $-\frac{1}{3}e$ or $-\frac{2}{3}e$? These subjects are discussed in the follow up Chapters.

2.4   Self-Mass (Self-Energy) of the Intrinsic Electron

In pre-AQM particle physics, the origin of self-mass (self-energy) for the electron has never been satisfactorily explained. As Pauli stated in 1945, "We will be considered the generation that left behind unsolved such essential problems as the electron self-energy." The explanation of the origin of electron self-mass on the basis of Higgs mechanism is a sign of desperation.

In AQM, under the General Compton Conditions, the c-ring electrostatic energy E_E and magnetostatic energy E_H are equal:

$$E_E = E_H \qquad (2\text{-}4)$$

The total self-energy and self-mass of the intrinsic electron are:

$$E_{\hat{e}} = E_E + E_H \qquad (2\text{-}5)$$
$$m_{\hat{e}} = m_E + m_H \qquad (2\text{-}6)$$

Here is the basic formula for total physical self-energy and self-mass of the intrinsic electron:

$$E_{\hat{e}} = 2\hbar\omega_c, \qquad (2\text{-}7)$$

and

$$m_{\hat{e}} = 2\hbar\omega_c / c^2 \qquad (2\text{-}8)$$

where ω_c is a classical Compton angular velocity of the c-ring.

Relations (2-7) and (2-8) provide the complete explanation for the c-ring physical self-energy and the origin of self-mass. *The intrinsic electron self-mass is 100% electromagnetism and can be computed, as explained in Section 2.10.*

No Higgs mechanism is required.

2.5   Spin of the Intrinsic Electron

In the pre-AQM particle physics, a value of electron spin is found experimentally and assigned to the electron as quantum parameter *by proclamation*. AQM provides a straightforward calculation of the value of intrinsic electron spin

Here is a classical formula for spin calculation:

$$S = m \upsilon r \qquad (2\text{-}9)$$

The physical properties of the c-ring can be expressed on the basis of classical electrodynamics. The c-ring electrostatic energy E_E does not contribute to spin. *Only magnetostatic energy E_H is the sole contributor to spin.*

♦ ♦ ♦

Postulate 48. *Spin of the intrinsic electron.*

The intrinsic electron spin is equal to $\hbar/2$. Only magnetostatic self-energy E_H or one half of its total self-energy $E_{\hat{e}}$ contributes to intrinsic electron spin.

$$S_{\hat{e}} = \hbar/2 \qquad (2\text{-}10)$$

This is AQM explanation of why fermions have spin $\hbar/2$.

In the case of bosons, the entire boson self-energy contributes to spin, resulting in spin ℏ.

♦ ♦ ♦

Postulate 48 explains the fermion spin ½ enigma. In SM spin ½ does not return to its original state after rotation of 360 degrees, but only upon a rotation of twice 360 degrees, namely, 720 degrees. There is no ontology in such explanation. In AQM, the electron Compton angular velocity is one half of the fermion velocity assumed in SM.

AQM explains that a classical Compton angular velocity of the c-ring produces only one half of fermion self-energy.

2.6 Magnetic Moment of the Intrinsic Electron

In classical electrodynamics, magnetic moment M is defined as $M = IA$, where I is total current and A is area. In the case of the c-ring:

$$I = \frac{e\omega_c}{2\pi} \text{ and } A = \pi R^2 = \frac{\pi c^2}{\omega_c^2} \qquad (2\text{-}11)$$

In terms of magnetostatic self-mass, the intrinsic electron magnetic moment is:

$$M = \tfrac{1}{4}\hbar \frac{e}{m_H}, \qquad (2\text{-}12)$$

or in terms of total self-mass m_e, the intrinsic electron moment is:

$$M = \frac{1}{2}\hbar\frac{e}{m_{\hat{e}}} = one\ Bohr\ magneton \qquad (2\text{-}13)$$

The intrinsic electron is the fundamental fermion of electromagnetic force. It does not have "anomaly" in its magnetic moment. In any case, the so-called "anomaly" is another SM fundamental misconception.

2.7 Intrinsic Electron Gyromagnetic Ratio. AQM Asserts that the Classical-Quantum Divide Does Not Exist

The gyromagnetic ratio is defined as M/S. In a classical example of the spinning cylinder with uniformly distributed electric charge Q on its surface and a total mass m, this is:

$$M/S = -Q/2m \qquad (2\text{-}14)$$

In the case of the intrinsic electron, the gyromagnetic ratio is (see 2-13):

$$M/S = -e/m_{\hat{e}} \qquad (2\text{-}15)$$

SM explains that discrepancy of factor two between (2-14) and (2-15) is an example of the classical-quantum divide. Such explanation is another misconception. So-called divide does not exist.

Here is the AQM explanation.

Intrinsic electron self-mass consists of two equal contributors: electrostatic self-mass m_E and magnetostatic self-mass m_H. Electrostatic self-mass does not contribute to spin or to magnetic moment. In this respect, it is a passive by-stander. Therefore, in formula (2-15) we should include only m_H:

$$M/S = -e/2m_H \qquad (2\text{-}16)$$

Comparing (2-14) and (2-16), one can see that in this particular case, the classical-quantum divide does not exist. Consideration of the g-factor is not required.

◆ ◆ ◆

Postulate 49. *Gyromagnetic ratio of intrinsic electron in both, classical and quantum, presentations.*

In classical presentation

$$M/S = -Q/2m \qquad (A).$$

In quantum presentation

$$M/S = -e/2m_H \qquad (B),$$

where $m_H = \hbar\omega_c/c^2$ is magnetostatic self-mass of the intrinsic electron. Only magnetostatic self-mass contributes to spin and magnetic moment of the intrinsic electron.

In this respect, the classical-quantum divide does not exist.

The gyromagnetic ratio (B) is valid for all intrinsic fermions representing fundamental forces: electromagnetism, weak, and strong.

A concept of g-factor becomes irrelevant.

◆ ◆ ◆

2.8   Stability of the Intrinsic Electron

There is an equilibrium of electrostatic and magnetostatic forces over the entire surface of the c-ring.

Magnetic field on outer surface is $B_0 = 0$. Magnetic field on inner surface is

$$B_i = \mu_0 I,$$ (2-17)

where $I = \dfrac{e\omega_c}{2\pi}$. Surface electric charge density is defined as:

$$\sigma = \frac{e}{2\pi r_c l_c}$$ (2-18)

where l_c is the length of the c-ring and $r_c = c/\omega_c$. Taking into consideration (2-17) and (2-18), we obtain magnetic field B_i on inner surface of c-ring

$$B = B_i = \mu_0 \sigma c$$ (2-19)

Electric field E on outer surface is:

$$E = \frac{\sigma}{\varepsilon_0}$$ (2-20)

Combining (2-19) and (2-20) we obtain

$$B = Ec$$ (2-21)

Outward pressure on the surface is caused by electrostatic field and equal to:

$$P_E = \sigma E$$ (2-22)

Inward pressure is caused by magnetostatic "pinch" and equal to:

$$P_B = -\sigma c E$$ (2-23)

Taking into consideration (2-22) and (2-23), we obtain equilibrium of electromagnetostatic forces on the surface:

$$P_B = -P_E$$ (2-24)

Tangential forces on the surface are also neutralized. All elements of electric charge are spinning in parallel to each other at the speed of light, thus neutralizing any repulsive or attractive tangential electric or magnetic forces.

At first glance, it appears that the intrinsic electron is stable. Electrostatic repulsive outward force applied to c-ring surface is balanced by magnetostatic inward pinch force over the entire surface of the c-ring.

This is only apparent stability. It appears that the c-ring is stable with any value of Compton radius and corresponding value of self-energy. That only means that c-ring is not stable at all. The intrinsic electron is looking for any opportunity to quickly release energy and create other inner structures. As the fundamental fermion, the intrinsic electron does not decay but cannot exist by itself in a stable state. In a specific pathway scenario, by releasing part of its self-energy within 10^{-22}-10^{-25} seconds, the intrinsic electron creates a neutrino-antineutrino pair, acquires a neutrino as a partner, releases antineutrino, and in combination with the neutrino, provides conditions for temporary stability and temporary lifetime, in cases of tau or muon, or acquires permanent stability and infinite lifetime in case of electron.

◆ ◆ ◆

Postulate 50. *The c-ring model for the fermions.*

In contrast to historical classical electrodynamic models, the c-ring model is mathematically accurate, requiring no approximation for all fermions.

◆ ◆ ◆

2.9   The Position Parameter of the Intrinsic Electron

The intrinsic electron consists of three components: the physical c-ring, the aphysical cylinder, and the elementary consciousness residing in the c-ring.

♦ ♦ ♦

Postulate 51. *The Position Parameter of the intrinsic electron*

The position of the c-ring along the aphysical cylinder axis is defined as the position parameter (PP), which is an irreducible quantum random parameter. There are no two intrinsic electrons with the same value of the position parameter.

♦ ♦ ♦

2.10   How to Calculate the Length of the C-ring for the Intrinsic Electron

The length of the c-ring of the intrinsic electron l_c is a fundamental constant. It belongs to the same class of fundamental constants as the speed of light c, the electric elementary charge e, the Planck constant h, and several others to be introduced and discussed in the follow up Chapters. By definition, the fundamental constants cannot be theoretically derived. They are preset by Nature and must be found experimentally.

As the principal contributor to electron self-mass, intrinsic electron self-mass is experimentally established and known as

$m_e = 0.511$ MeV/c^2 (less the neutrino self-mass). The constant l_c can be computed under the general Compton conditions. This is a 100% classical electrodynamics problem which can be handled by undergraduates and graduates. It would be their important contribution to foundational physics.

Here is the explanation. The c-ring geometry, the electron charge distribution, and the boundary conditions are well defined. To solve the problem, one has to select a specific value of Compton c-ring radius r_c, the corresponding value of Compton angular velocity ω_c, and the corresponding self-mass m_e. Here is a well known point: $r_c = 2 \times 3.86 \times 10^{-13}$ m, $\omega_c = 3.87 \times 10^{20}$ rad/sec, $m_e = 0.511$ MeV/c^2 (to be adjusted by neutrino self-mass) corresponding to $E_p = \hbar \omega_c \times 2$.

The solution of this problem requires computation of the magnetostatic and electrostatic field distributions, their energy density, and total intrinsic electron energy for various values of l_c. By extrapolation, one can find the correct value of constant l_c corresponding to correct value of self-mass. The correct value of l_c defines the correct value of intrinsic electron self-mass. All other values of l_c correspond to different values of self-mass and different values of Planck constant.

The length of the intrinsic electron c-ring is independent of the c-ring self-mass. One can select any value of Compton angular velocity with corresponding self-mass $m_{\hat{e}} = 2\hbar\omega_c / c^2$. Then computation would result in the same value of l_c as was found for "the known point". The length l_c is the same for the intrinsic electron, the intrinsic muon, and the intrinsic tau. Both, muon and tau represent higher energy excitation of the intrinsic electron as shown in Figure 12.

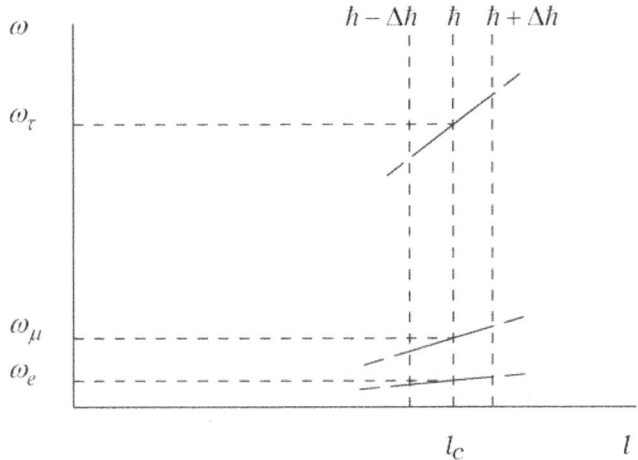

Figure 12
It is assumed that electric charge density is constant
(Chapter 6)

◆ ◆ ◆

Postulate 52. *The constants l_c and L_a are fundamental constants of the intrinsic electron.*

The value of the constant l_c is exact. Even a tiny deviation from the correct value of l_c would result in a deviation from the established value of the Planck constant and would create a conflict between the intrinsic electron and the photon.

◆ ◆ ◆

◆ ◆ ◆

Postulate 53. *AQM complete definition of the intrinsic electron.*

- *The intrinsic electron $\hat{e}^-$ is the fundamental fermion of electromagnetism. It is the exclusive carrier of negative electric charge, just as the intrinsic positron $\hat{e}^+$ is the exclusive carrier of positive electric charge. Both, $\hat{e}^-$ and $\hat{e}^+$ are antiparticles of each other.*

- *The intrinsic electron inner structure consists of the physical energy c-ring, the aphysical energy cylinder, and elementary consciousness residing in the c-ring.*

- *The surface of the intrinsic electron is made of a single elementary unit of negative electric charge (-e) with uniform charge density distribution.*

- *The intrinsic electron has two fields: electrostatic and magnetostatic. The fields are described by classical electrodynamics.*

- *The General Compton Conditions are especially applicable to the intrinsic electron: self-energy $E_p = \hbar\omega_c \times 2$, self-mass $m_{\hat{e}} = 2\hbar\omega_c/c^2$, electrostatic energy $E_E = \frac{1}{2}E_p$, magnetostatic energy $E_H = \frac{1}{2}E_p$, magnetic moment $M_{\hat{e}} = \dfrac{e\hbar}{4m_H}$, where $m_H = \frac{1}{2}m_{\hat{e}}$. Repulsive electrostatic force is balanced by magnetostatic pinch force over the entire c-ring surface.*

- *The intrinsic electron spin is equal to $\hbar/2$. Only its magnetostatic self-energy E_H or one half of its total self-energy $E_{\hat{e}}$ contributes to intrinsic electron spin.*

66

- *The aphysical General Compton Conditions are applicable to the aphysical cylinder as "imitation" of the physical General Compton Conditions with total aphysical energy $E_a = E_p/U$, where U is the universal constant.*

- *The length of the c-ring l_c and the length of the aphysical cylinder L_a are fundamental constants (to be determined).*

- *A free intrinsic electron expands radially with nearly the speed of light, releasing its energy in 10^{-22}-10^{-25} seconds, producing other inner structures, such as neutrino-antineutrino pairs, or quark-antiquark pairs.*

- *Along its pathway from high energy level to low energy level with intermediate energy releases, creating other inner structures, the intrinsic electron finally arrives at the ground energy level: the electron, where the intrinsic electron together with its partner, the electron neutrino, are trapped "forever" with the total self-energy of 0.511 MeV, including neutrino self-energy. During this process, helicity of the intrinsic electron is unchanged.*

- *There are no free intrinsic electrons in existence below the ground energy level (this statement is a subject to experimental verification).*

- *The AQM c-ring model of the intrinsic electron is mathematically accurate, requiring no approximation.*

- *Each individual intrinsic electron has a unique position parameter.*

- *Electric charge is a special state of matter, not yet recognized by science. By itself, electric charge has no self-energy, or gravitation, or inertia.*

♦ ♦ ♦

Chapter 3
The Intrinsic Neutrino and
The Fundamental Weak Force

3.1 Intrinsic Neutrino Definition

The intrinsic neutrino ($\hat{\nu}$) is the fundamental fermion of weak force. According to AQM, the intrinsic neutrino is the exclusive carrier of negative weak charge $-w$, just as the intrinsic antineutrino ($\hat{\bar{\nu}}$) is the exclusive carrier of positive weak charge $+w$.

3.2 Intrinsic Neutrino Inner Structure

The inner structure is shown in Figure 13. It consists of the physical c-ring, the aphysical cylinder (AC) and the elementary consciousness residing in the c-ring. It is my assumption that the intrinsic neutrino length is much shorter than the length of the intrinsic electron length ($l_c^\nu << l_c$).

3.3 Intrinsic Neutrino Field Configuration

The intrinsic neutrino c-ring surface is made up of negative weak charge $-w$ with uniform charge density distribution producing *neutrino-electrostatic field*. The intrinsic neutrino c-ring spins with Compton angular velocity generating *neutrino-magnetostatic field*.

The intrinsic neutrino field configuration is shown in Figure 14.

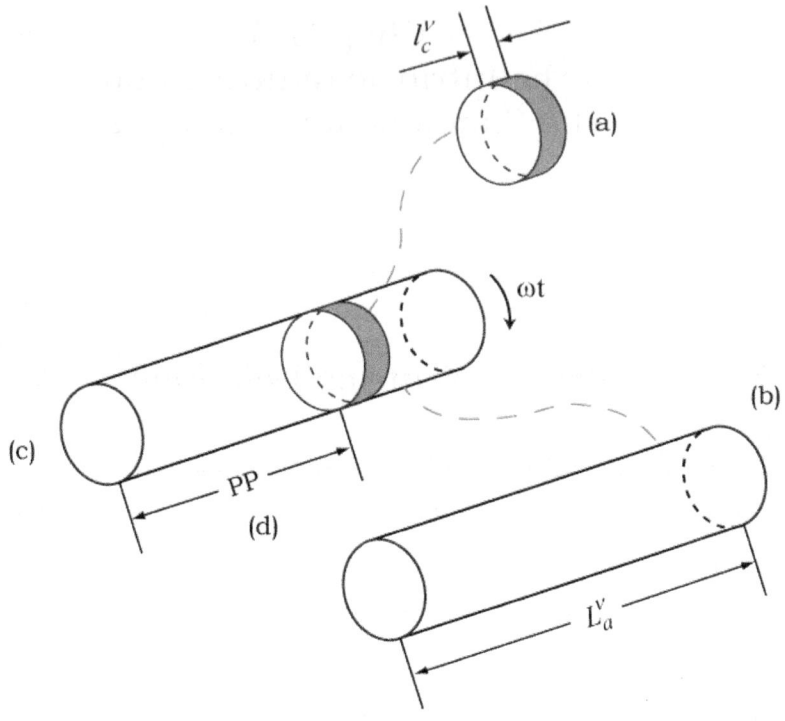

Figure 13

The intrinsic neutrino inner structure

    (a) physical c-ring

    (b) aphysical cylinder (AC)

    (c) neutrino inner structure

    (d) position parameter (PP)

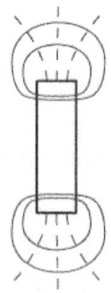

Figure 14

Intrinsic neutrino field configuration

70

According to AQM, the intrinsic neutrino- magnetostatic field and the intrinsic electron magnetostatic field have affinity, as explained later.

3.4   Intrinsic Neutrino Stability

Similar to the intrinsic electron, the intrinsic neutrino is not a stable elementary particle. It has tendency to release its self-energy by creating other inner structures provided it had sufficient self-energy.

3.5   Intrinsic Neutrino Self-Energy and Self-Mass

As explained in Chapter 6, each fundamental force has its own Planck constant. Weak force Planck constant is $\hbar_w = \beta\hbar$.

Intrinsic neutrino energy $E_{\hat{v}}$ is a total of neutrino electrostatic energy and neutrino magnetostatic energy, E_E and E_H, which are equal:

$$E_{\hat{v}} = E_E + E_H,$$

$$E_E = E_H,$$

$$E_{\hat{v}} = 2\,\hbar_w\omega_c,$$

where ω_c is classical Compton angular velocity of the c-ring. Self-mass is equal to

$$m_{\hat{v}} = 2\times\frac{\hbar_w\omega_c}{c^2},$$

where $m_H + m_E = m_{\hat{v}}$.

3.6   Intrinsic Neutrino Spin

The intrinsic neutrino is the fundamental fermion of weak electro-magnetism. Similar to the intrinsic electron, only one-half of intrinsic neutrino self-mass is applied to neutrino spin:

$$S_V = \frac{1}{2} \frac{E_{\hat{v}}}{c^2} c \frac{c}{\omega_c} = \frac{1}{2} \frac{E_{\hat{v}}}{\omega_c}$$

$$S_V = \frac{1}{2} \hbar_{\hat{v}}$$

SM assigns to neutrino spin of $\frac{1}{2}\hbar$, which is another fundamental misconception.

3.7   Intrinsic Neutrino Magnetic Moment and Gyromagnetic Ratio

The intrinsic neutrino magnetic moment M is equal to one weak Bohr magneton:

$$M_{\hat{v}} = \frac{w\hbar_w}{2m_{\hat{v}}}$$

Only one half of intrinsic neutrino self-mass is applied to intrinsic neutrino magnetic moment

Intrinsic neutrino gyromagnetic ratio is:

$$\frac{M_{\hat{v}}}{S_{\hat{v}}} = \frac{w}{2m_H}$$

◆ ◆ ◆

Postulate 54. *AQM complete definition of the intrinsic neutrino.*

- *The intrinsic neutrino ($\hat{v}$) is the fundamental fermion of weak electromagnetism. It is the exclusive carrier of negative weak charge −w, just as the intrinsic antineutrino ($\hat{\bar{v}}$) is the exclusive carrier of positive weak charge +w. Both, $\hat{v}$ and $\hat{\bar{v}}$ are antiparticles of each other.*

- *The intrinsic neutrino inner structure consists of the physical energy c-ring, the aphysical energy cylinder, and elementary consciousness residing in c-ring.*

- *The intrinsic neutrino c-ring is made up of negative weak electric charge with uniform charge density distribution.*

- *The intrinsic neutrino has two fields: neutrino electrostatic and neutrino magnetostatic. The fields are described by a classical weak electromagnetism.*

- *The intrinsic neutrino c-ring rotates with a classical Compton angular velocity*

$$\omega_c = c/r_c, \qquad (3\text{-}1)$$

where r_c is a classical Compton radius.

- *Radial forces at the c-ring surface are balanced. Outward repulsive electrostatic pressure P_o is opposed by equal inward magnetostatic pinch pressure P_i over the entire c-ring surface:*

$$P_o = -P_i. \qquad (3\text{-}2)$$

- *Under the General Compton Conditions, electrostatic energy E_o is equal to magnetostatic energy E_i:*

$$E_o = E_i \qquad (3\text{-}3)$$

73

- *There are no repulsive or attractive tangential forces along the c-ring surface. All elements of weak charge rotate in parallel to each other at the speed of light.*

- *The aphysical General Compton Conditions are applicable to aphysical cylinder as an "imitation" form of the physical General Compton Conditions with total aphysical energy $E_a=E_p/U$, where U is the universal constant.*

- *The length of the c-ring l_c^v and the length of the aphysical cylinder L_a^v are fundamental constants to be determined.*

- *Each individual intrinsic neutrino has a unique value of Position Parameter.*

◆ ◆ ◆

Chapter 4
The Duo-Electron: Definition, The Inner Structure, and Properties

4.1   Definition of the Duo-Electron

The duo-electron (symbol *e*) is a composite fermion of electromagnetism in duo-configuration, postulated and coined in AQM.

The duo-electron constituents are the intrinsic electron and the intrinsic positron with equal self-energies, opposite spins, and aligned magnetic momenta.

4.2   The Inner Structure of the Duo-Electron

The inner structure of the duo-electron consists of the intrinsic electron inner structure (-*e*) and the intrinsic positron inner structure (+*e*) with opposite helicities, as shown in Figure 15 (a).

Both inner structures, (-*e*) and (+*e*), are perfectly aligned, including their c-rings and aphysical cylinders which also means that both position parameters PP(-*e*) and PP(+*e*) are identical.

As explained in Chapter 9, such perfect alignment is a result of specifics in the origin and the formation of the duo-electron and the duo-positron. As pair of Majorana particles, the duo-electron and the duo-positron are two identical fermions of electromagnetism with opposite helicities.

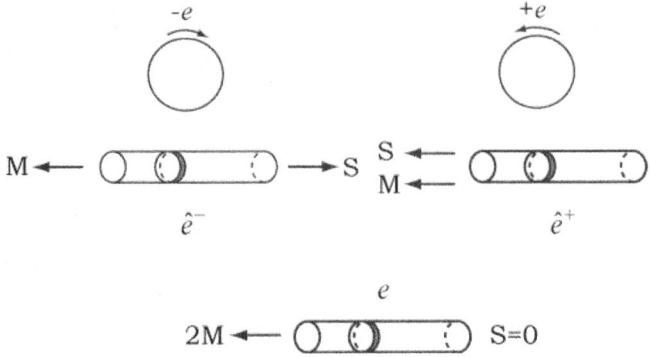

Figure 15 (a)

The duo-electron inner structure (e)

($\hat{e}^-$) – the inner structure of the intrinsic electron;

($\hat{e}^+$) – the inner structure of the intrinsic positron.

As shown in Figure 15 (b), both negative electric charge (-e) and positive electric charge (+e) rotate within zero space of the composite c-ring with opposite peripheral speed of light. Such dynamics prevents the annihilation of the composite c-ring.

This is one of AQM fundamental scientific discoveries.

4.3 Field Configuration and Balance of Opposing Magnetostatic Forces

The duo-electron inner structure has perfect alignment of both c-rings and, respectively, of both aphysical cylinders. As a result, the electromagnetic field configuration is transformed into something we have never seen in classical electrodynamics. Maxwell equations and special relativity are subjected to significant tension.

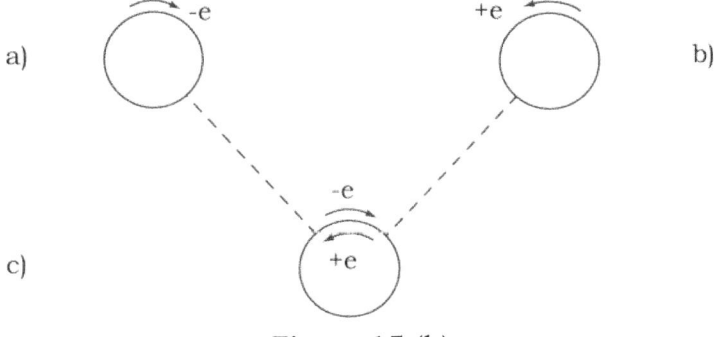

a)

b)

c)

Figure 15 (b)

a) cross-section of the c-ring of the intrinsic electron;

b) cross-section of the c-ring of the intrinsic positron;

c) cross-section of the composite c-ring of the duo-electron

(or the duo-positron).

In the system of two c-rings with opposite electric charges, opposite Compton angular velocities ω_c and $-\omega_c$ and opposite peripheral linear velocities $v = c$ and $v = -c$, electrostatic field is cancelled and subsequently, electrostatic energy is zero. The duo-electron is a single field fermion with magnetostatic field only.

The spinning c-rings produce a longitudinal magnetostatic field compressed on both sides of c-ring surface with a radial thickness $\delta = 0$ (see Figures 16 a, b). Both, inside and outside magnetostatic fields B_o and B_i are opposed to each other and equal in strength at the surface:

$$B_o = -B_i \qquad (4\text{-}1)$$

Radial forces at the c-ring surface are balanced. Outward repulsive magnetostatic pinch pressure P_o is opposed by equal inward magnetostatic pinch pressure P_i over the entire c-ring surface:

$$P_o = -P_i. \qquad (4\text{-}2)$$

Under the General Compton Conditions, outside magnetostatic energy E_o is equal to inside magnetostatic energy E_i:

$$E_o = E_i \qquad (4\text{-}3)$$

Total self-energy is equal:

$$E = E_o + E_i \qquad (4\text{-}4)$$

There are no repulsive or attractive tangential forces along c-ring surface. All elements of electric charge are spinning in parallel to each other at the speed of light.

The aphysical General Compton Conditions are applicable to the aphysical cylinder as an "imitation" of the physical General Compton Conditions with total aphysical energy $E_a = E_p/U$, where U is the universal constant.

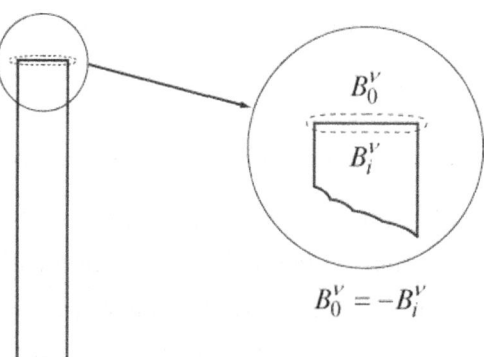

Figure 16 (a)

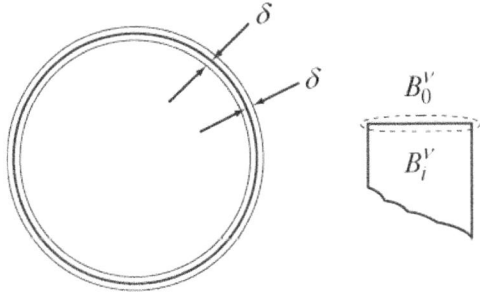

Figure 16 (b)

4.4   Duo-Electron Self-Corset Force

One of the most important properties of the duo-electron is "self-corset" force.

Although as I stated previously that magnetostatic inward and outward pressures P_i and P_o are balanced and magnetostatic fields B_i and B_o are equal in magnitude and confined to zero δ space, in fact, it is only an extremely good approximation but not absolute.

In reality, magnetostatic pressures are not absolutely equal. Inward magnetostatic pressure P_i prevails over outward pressure P_o.

Under such conditions, the strong self-corset binding force emerges, which reduces almost instantaneously classical Compton radius r_c acquired at the duo-electron formation, to quantum Compton radius r_q by many orders of magnitude.

The duo-electron transforms itself radially into an extremely small and extremely stable particle, perhaps even more stable than the electron.

4.5   Duo-Electron Basic Properties

- The duo-electron is extremely stable.

- The duo-electron is a single field fermion of electromagnetism with magnetostatic field only.

- Electrostatic self-energy is zero.

- Spin is zero (there are two equal and opposing ½ - inner spins).

- Magnetic moment is equal to two times of Bohr magneton.

- Self-corset force radially squeezes duo-electron from classical Compton radius r_c to quantum Compton radius r_q by many orders of magnitude.

- The duo-electron has very small cross-section of interaction.

- Gyromagnetic ratio $S/M = 0$.

A classical analogy is a system of two classical cylinders of equal geometry and equal mass, charged with equal but opposite charges and spinning with equal but opposite angular velocities.

Here is another example illustrating that classical-quantum divide does not exist.

The duo-electron has not yet been identified or discovered. Perhaps, it is misidentified. *One should expect a whole new class of duo-electrons in existence.*

4.6 Evolution of Duo-Electron Field Configuration as a theoretical and computational problem for undergraduates and graduates

One can compute several of electrostatic and magnetostatic field configurations of evolution of the duo-electron to the intrinsic electron by incrementally reducing positive electric charge in the intrinsic positron c-ring. This would be a challenging and rewarding task in foundational physics for undergraduates and graduates.

◆ ◆ ◆

Postulate 55. *The duo-electron. Complete definition and properties.*

- *The duo-electron is a composite fermion of electromagnetism, defined as a composite of intrinsic electron and intrinsic positron of equal self-energy with equal opposing ½ - spins and aligned magnetic momenta. Both physical c-rings and both aphysical cylinders are perfectly aligned, which is explained by the origin of the duo-electron.*
- *The duo-electron is a stable fermion.*
- *The duo-electron is a single field fermion with magnetostatic field only.*
- *The duo-electron has spin zero ($\frac{1}{2}\hbar - \frac{1}{2}\hbar$) and magnetic moment equal to two times of one Bohr magneton.*
- *Just prior the duo-electron formation and the beginning of the corset action, the duo-electron c-rings rotate with opposite Compton angular velocities,*

$$\omega_c = c/r_c \, ,$$

where r_c is a classical Compton radius.

- Spinning c-rings produce a longitudinal magnetostatic field compressed on both sides of the c-rings surface within space $\delta = 0$.

- Both, inside and outside, magnetostatic fields B_o and B_i are opposing and equal in strength at the surface:

$$B_o = -B_i$$

- Radial forces at the c-ring surface are balanced. Outward repulsive magnetostatic pressure P_o is opposed by equal inward magnetostatic pressure P_i over the entire c-ring surface:

$$P_o = -P_i.$$

- Under the General Compton Conditions, outside magnetostatic energy E_o is equal to inside magnetostatic energy E_i:

$$E_o = E_i$$

Total self-energy is equal to:

$$E = E_o + E_i$$

- There are no repulsive or attractive tangential forces along the c-ring surface. All elements of electric charge rotate in parallel to each other at the speed of light.

- In reality, inward magnetostatic pressure P_i prevails over outward magnetostatic pressure P_o .

Under such conditions, a strong self-corset binding force emerges, which reduces classical Compton radius r_c acquired at the start of the duo-electron formation, to quantum Compton radius r_q by more than ten orders of magnitude.

The duo-electron transforms itself radially into extremely stable particle. Similar to the neutrino, the duo-electron has very small cross-section of interaction.

- *The system of two c-rings with opposite but equal electric charges, spinning in mutual zero space with opposite but equal Compton angular velocities and opposite but equal peripheral velocities (the speed of light) as shown in Figures 15 (a, b) prevent c-rings from mutual annihilation. It is a unique property of particles with duo-configuration.*

- *The aphysical General Compton Conditions are applicable to aphysical cylinder as an "imitation" of the physical General Compton Conditions with total aphysical energy $E_a = E_p/U$, where U is a universal constant.*

- *The duo-electron and duo-positron are the same particles with opposite helicities.*

- *A symbol for the duo-electron is e.*

♦ ♦ ♦

Chapter 5
The Duo-Neutrino: Definition, The Inner Structure, and Properties

5.1   Definition of the Duo-Neutrino

The duo-neutrino (symbol ν) is a composite fermion of weak electromagnetism in duo-configuration, postulated and coined in AQM.

The duo-neutrino constituents are the intrinsic neutrino and the intrinsic antineutrino with equal self-energies, opposite spins, and aligned magnetic momenta.

5.2   The Inner Structure of the Duo-Neutrino

The inner structure of the duo-neutrino consists of the intrinsic neutrino inner structure ($\hat{\nu}$) and the intrinsic antineutrino inner structure ($\hat{\bar{\nu}}$) with opposite helicities, as shown in Figure 17 (a).

Both inner structures, ($\hat{\nu}$) and ($\hat{\bar{\nu}}$), are perfectly aligned, including their c-rings and aphysical cylinders which also means that both position parameters PP($\hat{\nu}$) and PP($\hat{\bar{\nu}}$) are identical.

As explained in Chapter 9, such perfect alignment is a result of specifics in the origin and the formation of the duo-neutrino and the duo-antineutrino. As pair of Majorana particles, the duo-neutrino and the duo-antineutrino are two identical fermions of weak electromagnetism with opposite helicities.

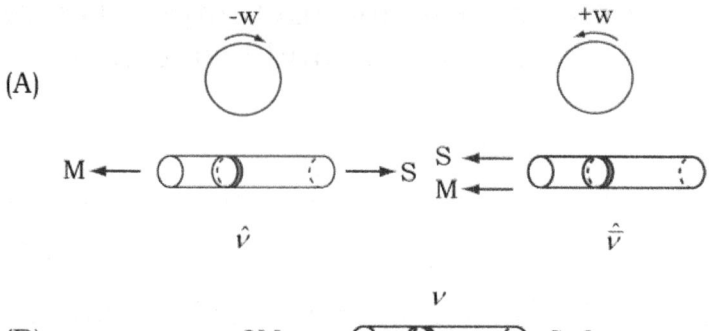

(A)

(B)

Figure 17 (a)

The duo-neutrino inner structure (v)

($\hat{v}$) – the inner structure of the intrinsic neutrino;

($\hat{\bar{v}}$) – the inner structure of the intrinsic antineutrino.

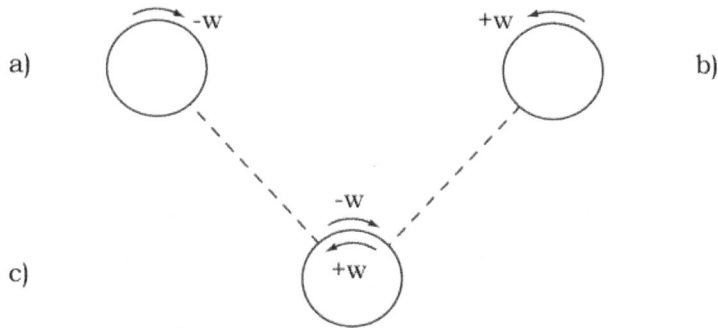

a)

b)

c)

Figure 17 (b)

a) cross-section of the c-ring of the intrinsic neutrino;

b) cross-section of the c-ring of the intrinsic antineutrino;

c) cross-section of the composite c-ring of the duo-neutrino

(or the duo-antineutrino).

As shown in Figure 17 (b), both negative weak electric charge ($-w$) and weak positive electric charge ($+w$) rotate within zero space of the composite c-ring with opposite peripheral speed of light. Such dynamics prevents the annihilation of the composite c-ring.

This is another AQM fundamental scientific discoveries.

5.3   Field Configuration and Balance of Opposing Magnetostatic Forces

The duo-neutrino inner structure has perfect alignment of both c-rings and, respectively, of both aphysical cylinders. As a result, the weak electromagnetic field configuration is transformed into something we have never seen in classical electrodynamics. Maxwell equations and special relativity are subjected to significant tension.

In the system of two c-rings with opposite electric charges, opposite Compton angular velocities ω_c and $-\omega_c$ and opposite peripheral linear velocities $v = c$ and $v = -c$, electrostatic field is cancelled and subsequently, electrostatic energy is zero. The duo-neutrino is a single field fermion with weak magnetostatic field only.

The spinning c-rings produce a longitudinal weak magnetostatic field compressed on both sides of c-ring surface with a radial thickness $\delta = 0$ (see Figures 18 a, b). Both, inside and outside weak magnetostatic fields B_o and B_i are opposed to each other and equal in strength at the surface:

$$B_o = -B_i \qquad (5\text{-}1)$$

Radial forces at the c-ring surface are balanced. Outward repulsive weak magnetostatic pinch pressure P_o is opposed by equal inward weak magnetostatic pinch pressure P_i over the entire c-ring surface:

$$P_o = -P_i. \qquad (5\text{-}2)$$

Under the General Compton Conditions, outside weak magnetostatic energy E_o is equal to inside weak magnetostatic energy E_i:

$$E_o = E_i \qquad (5\text{-}3)$$

Total self-energy is equal:

$$E = E_o + E_i \qquad (5\text{-}4)$$

There are no repulsive or attractive tangential forces along c-ring surface. All elements of weak electric charge are spinning in parallel to each other at the speed of light.

The aphysical General Compton Conditions are applicable to the aphysical cylinder as an "imitation" of the physical General Compton Conditions with total aphysical energy $E_a = E_p/U$, where U is the universal constant.

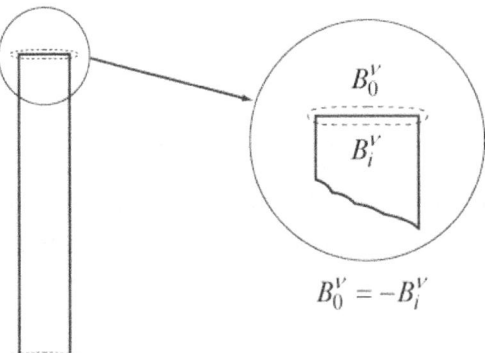

Figure 18 (a)

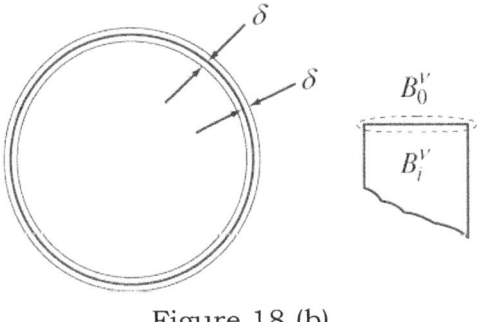

Figure 18 (b)

5.4   Duo-Neutrino Self-Corset Force

One of the most important properties of the duo-neutrino is "self-corset" force.

Although as I stated previously that weak magnetostatic inward and outward pressures P_i and P_o are balanced and weak magnetostatic fields B_i and B_o are equal in magnitude and confined to zero δ space, in fact, it is only an extremely good approximation but not absolute.

In reality, weak magnetostatic pressures are not absolutely equal. Inward weak magnetostatic pressure P_i prevails over outward pressure P_o.

Under such conditions, the strong self-corset binding force emerges, which reduces almost instantaneously classical Compton radius r_c acquired at the duo-neutrino formation to quantum Compton radius r_q by many orders of magnitude.

The duo-neutrino transforms itself radially into an extremely small and extremely stable particle.

5.5   Duo-Neutrino Basic Properties

- The duo-neutrino is extremely stable.

- The duo-neutrino is a single field fermion of weak electromagnetism with weak magnetostatic field only.

- Weak electrostatic self-energy is zero.

- Spin is zero (there are two equal and opposing $\frac{1}{2}\, \hbar_w$ inner spins).

- Magnetic moment is equal to two times of weak Bohr magneton.

- Self-corset force radially squeezes duo-neutrino from classical Compton radius r_c to quantum Compton radius r_q by many orders of magnitude.

- The duo-neutrino has very small cross-section of interaction.

- Gyromagnetic ratio $S/M = 0$.

A classical analogy is a system of two classical cylinders of equal geometry and equal mass, charged with equal but opposite charges, and spinning with equal but opposite angular velocities.

Here is another example illustrating that classical-quantum divide does not exist.

5.6   Evolution of Duo-neutrino Field Configuration as a theoretical and computational problem for undergraduates and graduates

One can compute several weak electrostatic and weak magnetostatic field configurations of evolution of the duo-neutrino to the intrinsic neutrino by incrementally reducing positive weak electric charge in the intrinsic antineutrino c-ring. This would be a challenging and rewarding task in foundational physics for undergraduates and graduates.

◆ ◆ ◆

Postulate 56. *The duo-neutrino. Complete definition and properties.*

- *The duo-neutrino is a composite fermion of weak electromagnetism, defined as a composite of intrinsic neutrino and intrinsic antineutrino of equal self-energy with equal opposing ½ $\hbar_w$ spins and aligned magnetic momenta. Both physical c-rings and both aphysical cylinders are perfectly aligned, which is explained by the origin of the duo-neutrino.*

- *The duo-neutrino is a stable fermion.*

- *The duo-neutrino is a single weak field fermion with weak magnetostatic field only.*

- *The duo-neutrino has spin zero (½ $\hbar_w$ - ½ $\hbar_w$) and magnetic moment equal to two times of weak Bohr magneton.*

- *Just prior the duo-neutrino formation and the beginning of the corset action, the duo-neutrino c-rings rotate with opposite Compton angular velocities,*

$$\omega_c = c/r_c ,$$

where r_c is a classical Compton radius.

- *Spinning c-rings produce a longitudinal weak magnetostatic field compressed on both sides of the c-rings surface within space $\delta = 0$.*

- *Both, inside and outside, weak magnetostatic fields B_o and B_i are opposing and equal in strength at the surface:*

$$B_o = -B_i$$

- *Radial forces at the c-ring surface are balanced. Outward repulsive weak magnetostatic pressure P_o is opposed by equal inward weak magnetostatic pressure P_i over the entire c-ring surface:*

$$P_o = -P_i.$$

- *Under the General Compton Conditions, outside weak magnetostatic energy E_o is equal to inside weak magnetostatic energy E_i:*

$$E_o = E_i$$

Total self-energy is equal to:

$$E = E_o + E_i$$

- *There are no repulsive or attractive tangential forces along the c-ring surface. All elements of weak electric charge rotate in parallel to each other at the speed of light.*

- *In reality, inward weak magnetostatic pressure P_i prevails over outward weak magnetostatic pressure P_o .*

Under such conditions, a strong self-corset binding force emerges, which reduces classical Compton radius r_c acquired at the start of the duo-neutrino formation, to quantum Compton radius r_q by many orders of magnitude.

The duo-neutrino transforms itself radially into extremely stable particle. The duo-neutrino has very small cross-section of interaction.

- *The system of two c-rings with opposite but equal weak electric charges, spinning in mutual zero space with opposite but equal Compton angular velocities and opposite but equal pe-*

ripheral velocities (the speed of light) as shown in Figures 17 (a, b) prevent c-rings from mutual annihilation. It is a unique property of particles with duo-configuration.

- *The aphysical General Compton Conditions are applicable to aphysical cylinder as an "imitation" of the physical General Compton Conditions with total aphysical energy $E_a = E_p/U$, where U is a universal constant.*

- *The duo-neutrino and duo-antineutrino are the same particles with opposite helicities.*

- *A symbol for the duo-neutrino is v .*

♦ ♦ ♦

Chapter 6
Weak Fundamental Force as a Branch of Electromagnetism ("Weak Electromagnetism")

6.1   Fractionation of the Intrinsic Electron C-ring as a Conceptual Exercise

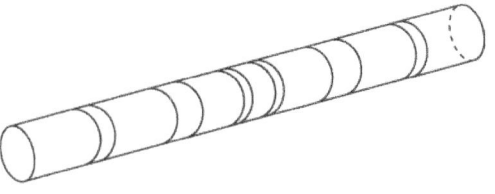

Figure 19

The intrinsic electron c-ring is shown in Figure 19. It is divided conceptually on n sections with random length Δl_i or normalized section length $\beta_i = \dfrac{\Delta l_i}{l_c}$, where l_c is the c-ring total length.

Each section can be presented independently with all its properties and parameters, as follows:

	Individual section	Total c-ring
Electric charge	$\Delta e_i = \beta_i\, e$	$e = \sum \Delta e_i$
Planck constant	$\Delta \hbar_i = \beta_i\, \hbar$	$\hbar = \sum \Delta \hbar_i$
Spin	$\Delta S_i = \beta_i\, S$	$S = \sum \Delta S_i$
Self-energy	$\Delta E_i = \beta_i\, E$	$E = \sum \Delta E_i$
Self-mass	$\Delta m_i = \beta_i\, m$	$m = \sum \Delta m_i$
Magnetic moment	$\Delta M_i = \beta_i\, M$	$M = \sum \Delta M_i$

Each section can be presented as a standalone c-ring of a mini-intrinsic electron.

Nature has decided that outside the nucleons there must be only two intrinsic electrons: one with electric charge e for electromagnetism and another one with weak electric charge w for weak electromagnetism.

Therefore, we have a great opportunity to find long-standing properties for weak electromagnetism as related to neutrino properties, such as weak electric charge, weak Planck constant, weak spin, weak self-energy, weak self-mass, and weak magnetic moment.

6.2   Fundamental Weak Force is a Branch of Electromagnetism

Here is another observation. At the fundamental level, Nature does not tolerate redundancy. But, here we have an obvious redundancy – both electromagnetism and weak fundamental force are parallel.

The redundancy can be resolved by postulating that fundamental weak force is a branch of electromagnetism (weak electromagnetism).

If this is the case, then the name "neutrino" is a misnomer. A more appropriate name would be "electrino". However, the historical tradition prevails leaving me no choice but to accept the established name "neutrino". However, I will use "electrino" once in a while where I think it is appropriate.

If we know parameter β, then we would be able to determine all above properties and parameters, including neutrino self-mass which remains a long-standing problem in fundamental physics. According to SM, neutrino self-mass is zero which is another SM fundamental misconception. However, as shown in Chapter 10, we are very fortunate – we know value of parameter β with great precision obtained from experiments in 1987 with a Penning trap [5]. My greatest respect and gratitude go to experimental physicists for providing data of such enormous fundamental significance.

6.3 Comparison of Electromagnetism with Weak Electromagnetism

Table 6.3-A

	Intrinsic Electron	Intrinsic neutrino in weak force	Intrinsic neutrino in weak electro-magnetism
C-ring length	l_c	l_c^v	l_c^v
l_c^v / l_c	-	-	β
Charge	e	w	βe
Plank constant	$\hbar$	$\hbar_w$	$\beta\hbar$
Spin	$\frac{1}{2}\hbar$	$\frac{1}{2}\hbar_w$	$\frac{1}{2}\beta\hbar$
Self-energy	$2\hbar\omega_c$	$2\hbar_w\omega_c$	$2\beta\hbar\omega_c$
Self-mass	$m_{\hat{e}} = \dfrac{2\hbar\omega_c}{c^2}$	$m_{\hat{v}} = \dfrac{2\hbar_w\omega_c}{c^2}$	$m_{\hat{v}} = \dfrac{2\beta\hbar\omega_c}{c^2}$
Magnetic moment	$\dfrac{e\hbar}{2m_{\hat{e}}}$	$\dfrac{w\hbar_w}{2m_{\hat{v}}}$	$\dfrac{\beta^2 e\hbar}{2m_{\hat{v}}}$

Table 6.3-B

	Intrinsic Electron	Electron neutrino	Electron
Charge	$-e$	0	$-e$
Plank constant	$\hbar$	$\hbar_w$	$\hbar, \hbar_w$
Spin	$\frac{1}{2}\hbar$	0	$\frac{1}{2}\hbar$
Self-energy	$E(\hat{e}^-) = 2\hbar\omega_c$	$2\beta\hbar\omega_c$	$E(\hat{e}^-)(1+\beta)$
Self-mass	$m(\hat{e}^-) = \dfrac{2\hbar\omega_c}{c^2}$	$\beta\, m(\hat{e}^-)$	$m(\hat{e}^-)(1+\beta)$
Magnetic moment	$M(\hat{e}^-) = \dfrac{e\hbar}{m(\hat{e}^-)}$	$2\beta\, M(\hat{e}^-)$	$M(\hat{e}^-)(1+2\beta)$

◆ ◆ ◆

Postulate 57. *Intrinsic neutrino - properties and parameters.*

- *Fundamental weak force is a branch of electromagnetism, coined in AQM as weak electromagnetism.*

- *Plunck constant of fundamental weak force,*

$$\hbar_w = \beta\hbar .$$

- *Intrinsic neutrino spin,*

$$S = \frac{1}{2}\,\beta\hbar .$$

- *Intrinsic neutrino elementary charge unit in terms of electron elementary electric charge,*

$$w = \beta e .$$

- *Intrinsic electron neutrino self-mass,*

$$m_{\hat{v}} = 2\frac{\hbar_w \omega_c}{c^2} .$$

- *Intrinsic neutrino magnetic moment,*

$$M_{\hat{v}} = \frac{w\hbar_w}{2m_{\hat{v}}} .$$

◆ ◆ ◆

Chapter 7
The Phontino as the Photon of Weak Electromagnetism.
The Inner Structure and Properties.

Here I postulate existence of the photon of weak electromagnetism (coined as *the phontino*). The phontino, γ_n, is the mediator of weak electromagnetism interaction. It is the quantum of *inverted* weak electromagnetostatic energy. As a boson, it carries no charge. It travels at the speed of light in free space. The phontino has self–mass in its system of self-reference $v = c$. In analogy with the photon, phontino is stable.

The phontino inner structure is shown in Figure 20. It consists of the physical c-ring, the aphysical cylinder, and the elementary consciousness residing in the c-ring.

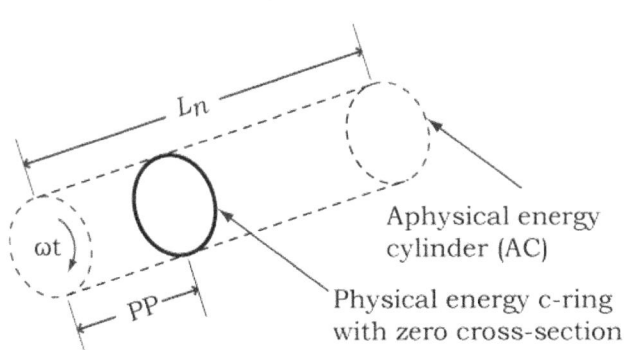

Figure 20

The inner structure of the phontino

The phontino c-ring is a spinning bundle of inverted weak electromagnetostatic energy with Compton angular velocity $\omega = c/r$. The phontino c-ring has zero cross-section. Its inverted weak electromagnetostatic energy is distributed uniformly along the c-ring circumference.

The phontino has spin $S = \hbar_w$ in the direction of travel or $S = -\hbar_w$ in the direction opposite to travel. Two helicities, right-handed and left-handed, correspond to two directions of rotation of the c-ring energy bundle.

As in the case of the photon, under certain conditions, the phontino energy bundle splits in two opposite rotating sub-bundles with the energy split at any ratio.

As a result, phontino spin can be found with any value in the range $[-\hbar_w, +\hbar_w]$, including a case of spin zero ("hidden" spin) where two sub-bundles are equally split and two halves spins opposing each other.

Existence of the phontino has yet to be confirmed.

The annihilation (the transformation) of intrinsic neutrino and intrinsic antineutrino produces two phontinos:

$$\hat{\nu} + \hat{\bar{\nu}} = \gamma_n + \gamma_n$$

The phontino energy, momentum, wavelength, and quantum radius are presented as:

$$E = \hbar_w \omega = h_w \nu = \lambda_w \, c/\lambda ,$$

$$\bar{p} = \hbar_w \bar{k} ,$$

where

$$k = |\bar{k}| = 2\pi/\lambda ;$$

$$p = h_w \nu/c = h\omega/\lambda ,$$

and

$$r = \frac{\lambda}{2\pi} = \frac{c}{\omega}$$

The Planck constant for weak electromagnetism is equal to

$$h_w = \beta\, h.$$

The normalized length of phontino aphysical cylinder L_n/λ_n is a fundamental aphysical constant.

Same as the photon, the phontino is stable in free space for it has no pathways for transferring its energy.

♦ ♦ ♦

Postulate 58. *The phontino as the photon of weak electromagnetism. The inner structure and properties.*

The phontino is the photon of weak electromagnetism and the mediator of weak electromagnetism interactions. The phontino is the quantum of inverted weak electromagnetostatic energy. It carries no charge. It travels at the speed of light in free space.

The phontino inner structure consists of the physical energy c-ring, the aphysical energy cylinder, and the elementary consciousness residing in the c-ring. A location of the c-ring along the aphysical cylinder axis is defined as Position Parameter, unique to each individual phontino.

The phontino inner structure spins at classical Compton angular velocity ω_c with Compton radius r_c. Phontino spin S can have any value in the range of $[-\hbar_w, +\hbar_w]$.

Phontino physical field is an inverted weak electromagnetostatic field. Phontino physical energy is an inverted weak electromagnetostatic energy. The phontino c-ring has zero cross-section and consists of the bundle of inverted weak electromagnetostatic energy spinning at Compton angular velocity. In some situations, the bundle splits into two sub-bundles at any ratio, spinning with opposite Compton angular velocities.

The phontino has two quantum states, the full-fledged state and the self-entangled state. In case of the latter, the phontino splits on several a-fractions, including the host. All a-fractions are connected to the host via c-links.

The aphysical cylinder imitates most of the c-ring properties, including aphysical spin and aphysical self-energy, with universal constant U observed.

The phontino self-energy E_n and self-mass m_n are defined as $\hbar_w \omega_c$ and $\hbar_w \omega_c / c^2$, where $\hbar_w$ is Planck constant of weak electromagnetism. The phontino is not a "massless" particle. Phontino self-mass m_n is defined in its system of self-reference $v=c$ as the c-ring spin energy, which is invariant in all systems of reference.

Phontino inverted weak electromagnetostatic self-energy density or self-mass density is uniformly distributed along the c-ring surface.

Similar to the photon, the phontino has no kinetic energy, unless it interacts with an "absorbing" entity. In such a case, the phontino spin energy is transformed into kinetic energy transferred with its spin to absorbing entity. The aphysical cylinder imitates this process.

The phontino c-ring has zero cross-section, therefore, the phontino does not interact with other phontinos or other bosons.

The constant L_n / λ_n is a fundamental aphysical constant.

The phontino is stable traveling in free space for it has no pathways for transferring its energy.

♦ ♦ ♦

Chapter 8
Three Separate Planck Constants, One for Each Fundamental Force: Electromagnetism, Weak Electromagnetism, and Strong Force.

The origin of Planck Constant h is electromagnetism.

In 1900, Max Planck successfully solved the problem of black body radiation on the basis of quantization of radiation energy. He has shown that statistically each "energy packet" is proportional to frequency,

$$E = hf$$

(or in contemporary terms, $E = \hbar\omega$, $\hbar = h/2\pi$).

Planck was able to calculate the value of h from experimental data of black body radiation spectrum. This was a first step toward development of quantum mechanics. It was also a giant step.

In 1905, Einstein extended such concept to the light by proposing that light consists of particles (quanta). Each quanta (photon) carries discreet energy proportional to its frequency υ

$$E = h\upsilon, \text{ or } E = hc/\lambda.$$

Using his approach, Einstein successfully explained photoelectric effect in 1909.

Then, in 1913, Niels Bohr introduced the first quantized model of the atom using Planck constant. The electron in a Bohr atom could only have certain defined orbits with energy $E(n)$ and orbital momentum $L(n)$,

$$E(n) = -hcR_e/n^2 \text{ and } L(n) = h_n, \quad n = 1, 2, 3, \dots$$

where R_e is the Rydberg constant.

As development of quantum mechanics progressed, Planck constant was showing up everywhere: in zero rest mass, elementary charge of the electron, Bohr magneton, the de Broglie relation, Schrödinger equation, Dirac equation, and the Heisenberg uncertainty principle. It was a triumph of the Planck constant.

It has been assumed that *the Planck constant is universal and applicable equally to each fundamental force of interaction.*

Looking back into history of quantum mechanics, one can make an astonishing discovery that, in fact, the Planck constant is always associated with electromagnetism. The extension to weak and strong interactions is done by *proclamation.*

I have discovered that using the Planck constant for the construction of the inner structures of composite elementary particles, such as leptons and quarks, creates insurmountable problems.

Therefore, I declare that the Planck constant is not universally applicable to each of three fundamental forces of interaction. It requires three separate Planck constants, one for each fundamental force: electromagnetism, weak electromagnetism, and strong force.

♦ ♦ ♦

Postulate 59. Three quantum dynamics theories.

There are three quantum dynamics theories, one for each funda-mental force: electromagnetism, weak electromagnetism, and strong force:

1. *quantum electrodynamics;*
2. *quantum weak electrodynamics; and*
3. *quantum strong force dynamics.*

♦ ♦ ♦

♦ ♦ ♦

Postulate 60. *Three types of Planck constant, one for each fundamental force.*

In the quantum reality, there are three separate Planck constants:

1. *The Planck constant h for electromagnetism. This is the established Planck constant h.*
2. *The Planck constant h_w for weak electrodynamics and weak interaction.*
3. *The Planck constant h_s for strong force and strong interaction.*

Based on relative strength of forces, here are relative values of Planck constants:

h is Planck constant for electromagnetism;

$h_w = \beta \times h$ for weak electromagnetism where

β = 0.000 579 826 094 or approximately 5.8×10⁻⁴ (as shown
in Chapter 10), and

$h_s = 100×h$ *for strong force (approximately).*

Three separate Planck constants with such vast difference in their
values are absolutely indispensable for the construction of the inner
structures of composite elementary particles, such as leptons and
quarks.

The precise value of h_s can only be established experimentally.

◆ ◆ ◆

Gravity is not really subject of my study. According to general
relativity, gravity defines space-time configuration. Some scientists
believe that quantum gravity leads to the quantization of space-time
domain. That would be a nightmarish scenario. Nature is not confused. Nature is elegant and majestic in its simplicity and sophisticated in its complexity.

In my opinion, gravity is not part of quantum reality.

If this were the case, then there would be a Planck constant for
gravity, h_g, with approximate value of $10^{-41}×h$. Bosonic graviton
would have spin $S = h_g/2$. In contrast, the string theory predicts a
value of spin for graviton of $S = 2\hbar$, which is an absurdity. Such value of spin would be more appropriate for strong force.

It is conceivable that there exist other fundamental forces beyond these three, such as force four, five, six, seven... These forces
are much more subtle and have yet to be discovered.

◆ ◆ ◆

Postulate 61. Three types of bosons, one for each fundamental force.

There are three types of bosons, mediators of fundamental forces, one for each force - electromagnetism, weak electromagnetism, and strong force:

1. *The photon with energy $E = \hbar\omega$ and spin $S = \hbar$;*

2. *The weak photon (the phontino) with energy $E = \hbar_w\omega$ and spin $S = \hbar_w$; and*

3. *The gluon with energy $E = \hbar_s\omega$ and spin $S = \hbar_s$.*

Bosons carry no charge.
Bosons travel with the speed of light in free space.

◆ ◆ ◆

In my definition, the gluon does not carry color and anticolor charges. The bosons carry no charge. The W gauge boson and the Higgs boson are SM fundamental misconceptions.

◆ ◆ ◆

Postulate 62. Three types of intrinsic fermions, one for each funda-mental force.

There are three types of intrinsic fermions, one for each fundamental force – electromagnetism, weak electromagnetism, and strong force:

1. The intrinsic electron with self-energy $E=2\hbar\omega_c$ and spin $S=\frac{1}{2}\hbar$.

   The intrinsic electron and the intrinsic positron are the exclusive carriers of elementary electric charge, negative or positive, respectively.

2. The intrinsic neutrino (the intrinsic electrino coined by me) with self-energy $E = 2\hbar_w\omega_c$ and spin $S = \frac{1}{2}\hbar_w$.

   The intrinsic neutrino and the intrinsic antineutrino are the exclusive carriers of elementary weak electric charge, negative ($-w$) or positive ($+w$), respectively.

3. The intrinsic fermion of strong force (not yet discovered) has energy $E = 2\hbar_s\omega$ and spin $S = \frac{1}{2}\hbar_s$. The intrinsic strong fermion and the intrinsic strong antifermion are the exclusive carriers of elementary strong (color) charge.

These are the basic fermions, the truly elementary particles.

They cannot be split further. This is the last inner structure frontier on the way down toward the smallest things.

♦ ♦ ♦

Postulate 63. *Three types of the de Broglie relation, one type for each fundamental force.*

The de Broglie relations are converted into three separate relations, one type for each fundamental force. Each type defines the length of a physical kinetic energy column:

1. *For electromagnetism,* $l_{ke} = h/|\overline{p}|$;

2. *For weak electromagnetism,* $l_{kw} = h_w/|\overline{p}|$;

3. *For strong fundamental force,* $l_{ks} = h_s/|\overline{p}|$.

It is remarkable that the Heisenberg uncertainty principle is invariant relative to the de Broglie relation transformation for all three fundamental forces.

♦ ♦ ♦

♦ ♦ ♦

Postulate 64. *Three types of the classical Maxwell electromagnetism, one type for each fundamental force.*

There are three types of the classical Maxwell electromagnetism, one type for each fundamental force:

1.   *For electromagnetism, it is the classical Maxwell electromagnetism.*

2.   *For weak electromagnetism, it is the classical Maxwell weak electromagnetism.*

3. For strong force, it is the classical Maxwell strong force electromagnetism.

♦ ♦ ♦

♦ ♦ ♦

Postulate 65. *Three types of Schrödinger equation, one for each fundamental force.*

Schrödinger equation is expanded into three separate equations, one type for each fundamental force.

♦ ♦ ♦

♦ ♦ ♦

Postulate 66. *Three types of Dirac equation, one for each intrinsic fermion of each force.*

Dirac equation is expanded into three separate equations, one for each intrinsic fermion (and intrinsic antifermion): the intrinsic electron, the intrinsic neutrino (the intrinsic electrino), and the intrinsic fermion of strong force (not discovered yet).

♦ ♦ ♦

Contrary to the established theoretical understanding, the Dirac equation does not describe the electron and its properties. The electron is a composite elementary particle with properties far beyond existing application of the Dirac equation. Insisting on the application of the Dirac equation to the description of the electron produces a fundamental misconception of the anomalous electron moment and misconception in g-factor.

Then, what is the Dirac equation? It is an equation for the description of intrinsic fermions (and intrinsic antifermions) such as the intrinsic electron for electromagnetism, the intrinsic neutrino for weak electromagnetism, and the intrinsic strong fermion for strong fundamental force.

◆ ◆ ◆

Postulate 67. *Three types of the Heisenberg uncertainty principle.*

There are three types of Heisenberg uncertainty principle, one type for each fundamental force:

1. *For electromagnetism, $|\Delta E \cdot \Delta t| \geq h$;*

2. *For weak electromagnetism, $|\Delta E \cdot \Delta t| \geq h_w$; and*

3. *For strong force, $|\Delta E \cdot \Delta t| \geq h_s$.*

The uncertainty is substantially reduced for weak electromagnetism.

Surprisingly, the Heisenberg uncertainty principle is not affected by introduction of three separate Planck constants, as long as it is applied consistently within each fundamental force domain.

115

On the other hand, it is not really surprising because AQM travels along a natural scientific trajectory.

♦ ♦ ♦

Chapter 9
The Electron – Two Visions: QM vs AQM

9.1   Introduction

From the time of its discovery in 1897, the electron has been one of the most studied elementary particles, both theoretically and experimentally. Scientific history of the electron is remarkable and dramatic. My work is not about the history of the electron.

One would think that since most of the properties of the electron have been discovered, there is not much left to discover. How mistaken such a notion is!

Like in the case of the photon, AQM dramatically expands scientific knowledge about the electron adding a plethora of new fundamental properties. In the fundamental understanding of the electron, there exists a surprisingly huge gap between the Standard Model and Aphysical Quantum Mechanics.

9.2   SM Understanding of the Electron

According to SM, the electron is the basic fermion of electromagnetism. It is generally thought to be a fundamental particle because it has no known constituents. It is assumed to be a point-like particle with a point-like electric charge, with no spatial extent and no inner structure.

The SM concept of a point-like electron conflicts with experimental observations in a Penning trap, which points to non-zero size of the electron with the upper limit of radius to be 10^{-22} meters [4].

Electron spin and magnetic moment are derived experimentally and assigned to the electron by proclamation as intrinsic quantum parameters. SM has no explanation for the origin of electron self-mass.

SM quantum model of the electron is described in Chapter 1. The model requires a creation of clouds of virtual particle-antiparticle pairs forming the vacuum polarization. Electron charge and mass are assumed to be point-like.

SM assigns wave properties to the electron based on the wave-particle duality, which are demonstrated in interference and diffraction experiments. Somehow, the physically indivisible electron is able to pass through two parallel slits simultaneously. This enigma remains unexplained in SM.

The electron is a fermion with spin $\hbar/2$. According to SM, all fermions have spin $\hbar/2$, which is another SM fundamental misconception.

The electron has a small anomaly in its magnetic moment explained theoretically as the result of infinite number of QED contributions from quantum fluctuations. The calculation is considered to be a triumph of QED.

According to quantum field theory, which is a part of SM, it would require infinite number of diagrams to arrive at a complete definition of the electron.

9.3   AQM Vision of the Electron

The AQM electron is a composite elementary particle consisting of the intrinsic electron and the electron neutrino (duo-electrino, coined by me). Each constituent has the inner structure consisting of the physical energy c-ring, the aphysical energy cylinder, and the elementary consciousness residing in the c-ring.

The intrinsic electron is the fundamental fermion of electromagnetism. The electron neutrino is the duo-fermion of weak electromagnetism, consisting of the intrinsic neutrino with negative weak electric charge (-w) and the intrinsic antineutrino with positive weak electric charge (+w) and opposite helicity. The electron neutrino has zero weak electric charge and is a single field fermion with magnetostatic field only.

The electron is the lepton at the ground state. Electron can be formed in variety of interactions. As an example, the electron formation from muon decay, shown in Figures 21 and 22, is presented in SM and AQM interpretations, respectively.

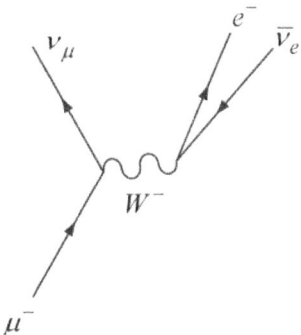

Figure 21

SM presentation of muon decay

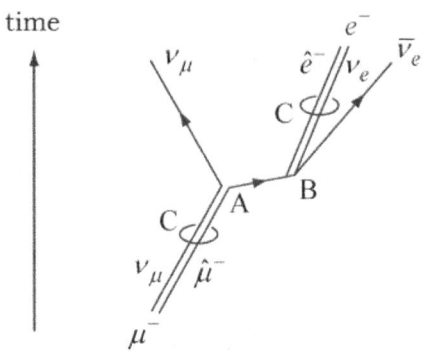

time

Figure 22

AQM presentation of muon decay and the electron formation
(for explanation, see Chapter 10)

One can notice that in contrast to Feynman diagrams, all ar-
rows are in the direction of time flow. Time always goes in one direction and no excuses are accepted.

The intermediate vector boson W shown in Figure 21 is another
SM fundamental misconception. Bosons carry no charge. To say
the boson W carries electric charge means that boson carries fermionic electron structure, which is an absurdity. For many decades
diagrams such as those shown in Figure 21 have been included in
particle physics books and in all particle physics textbooks. What is
presented as the intermediate vector boson is in fact the intrinsic
electron on its way to the electron formation. In this case the intrinsic electron has only one scenario – the formation of the electron in
its final quantum ground stage.

The electron has a finite non-zero size. Electron spin and magnetic moment are produced by *combined spinning of electric charge
and weak electric charge at Compton angular velocity.*

The electron has self-mass of 0.511 MeV/c² as a sum of electrostatic and magnetostatic energy of the intrinsic electron and magnetostatic energy of the electron neutrino. No Higgs particle, Higgs field

or Higgs mechanism is required to explain the origin of the electron self-mass.

The electron is not a wave. Its interference and diffraction properties are explained by its self-entanglement and aphysical-aphysical interaction of the host and a-fractions with periodic aphysical structures such as two-slit or Ni crystal.

In a case of two-slit experiment, the host with its physical substance goes intact through one slit and an aphysical a-fraction goes simultaneously through a second slit.

Here is the explanation for so-called "anomaly" in electron magnetic moment. Electron magnetic moment is superposition of intrinsic electron magnetic moment and electron neutrino magnetic moment. What is called "the anomaly" is in fact the magnetic moment of the electron neutrino. No vacuum fluctuations or pairs of virtual photons in the vacuum are required.

"The anomaly" is another SM fundamental misconception.

Chapter 10
AQM Theory of the Electron

10.1   The Electron Inner Structure

Surprisingly, the electron has hardly been explored in particle physics.

AQM describes in detail the electron inner structure and includes a plethora of new properties.

The electron e^- is a composite fermion of electromagnetism and weak electromagnetism consisting of the intrinsic electron $\hat{e}^-$ and the electron neutrino ν_e

$$e^- = \left\{ \begin{matrix} \hat{e}^- \\ \nu_e \end{matrix} \right\}$$

The description of electron neutrino is presented in Chapter 5. The electron neutrino inner structure has duo configuration consisting of the intrinsic electron neutrino $\hat{\nu}$ with negative weak electric charge $-w$ and the intrinsic electron antineutrino $\hat{\bar{\nu}}$ with positive weak electric charge $+w$ and opposite (reversed) helicity.

$$\nu_e = \left\{ \begin{matrix} \hat{\nu}_e \\ \hat{\bar{\nu}}_e^R \end{matrix} \right\}$$

The electron neutrino has zero weak electric charge $(-w+w=0)$ and zero spin. Its magnetic moment is equal to two times a weak Bohr magneton and is aligned with intrinsic electron magnetic moment.

The electron neutrino ν_e and the electron antineutrino $\bar{\nu}_e$ are Majorana particles. They are antiparticles to each other.

The entire inner structure of the electron e^- with its constituents, the intrinsic electron $\hat{e}^-$ and the electron neutrino ν_e, is shown in Figure 23 (a), (b), (c).

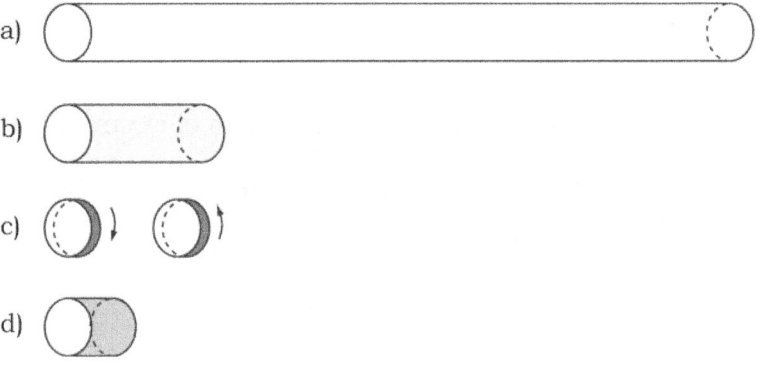

Figure 23 (a)

Constituents of the electron inner structure:
a) aphysical cylinder of intrinsic electron;
b) aphysical cylinder of electron neutrino;
c) c-rings of intrinsic neutrino and intrinsic antineutrino with opposite helicities;
d) c-ring of intrinsic electron.

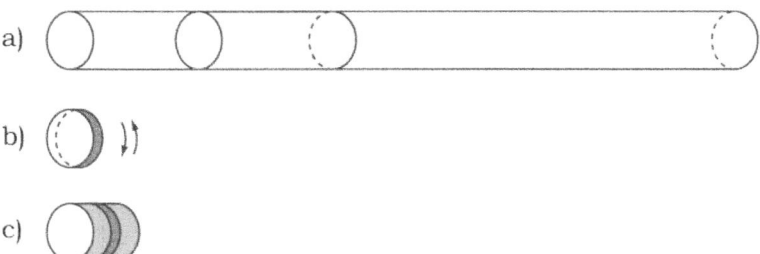

Figure 23 (b)

Sub-assemblies of the electron inner structure

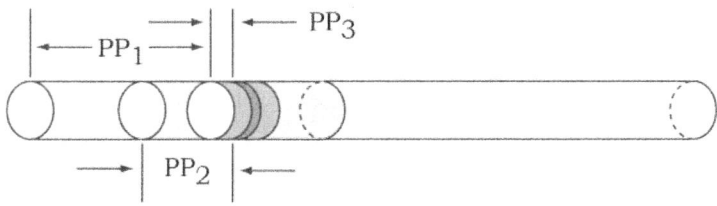

Figure 23 (c)

Complete inner structure of the electron

Each constituent of the electron, such as the intrinsic electron or the electron neutrino, consists of the physical c-ring and the aphysical cylinder spinning after the electron formation at quantum Compton radius and with quantum Compton angular velocity to be explained later.

The electron inner structure has three position parameters: PP_1, PP_2 and PP_3 (see Figure 23 (c)). PP_1 defines the position of the intrinsic electron c-ring relative to the aphysical cylinder; PP_2 defines the position of the neutrino c-ring relative to the neutrino aphysical cylinder; PP_3 defines the position of the neutrino c-ring relative to the intrinsic electron c-ring.

Position parameters are irreducible quantum random "hidden variables". Position Parameters make each electron unique. There are no two identical electrons.

One can find several different elementary particle interactions where the electron is formed. In the follow-up Sections 10.2, 10.4 and 10.5, I only consider an example of the electron formation in muon decay.

10.2   Electron Neutrino as a Corset

According to AQM, the neutrino plays a fundamental role in particle physics as the "corset" with a function of permanently trapping intrinsic electron, reducing it from classical Compton c-ring radius to quantum Compton c-ring radius, thus preventing it from expanding and releasing its energy below the ground energy level of 0.511 MeV. Intrinsic electrons with self-energy of less than 0.511 MeV do not exist.

Figure 23 (c) shows the inner structure of the full-fledged electron.

Figure 24 shows the intrinsic electron c-ring and the electron neutrino c-ring in the corset configuration prior the corset action. *Here the classical world meets the quantum world.*

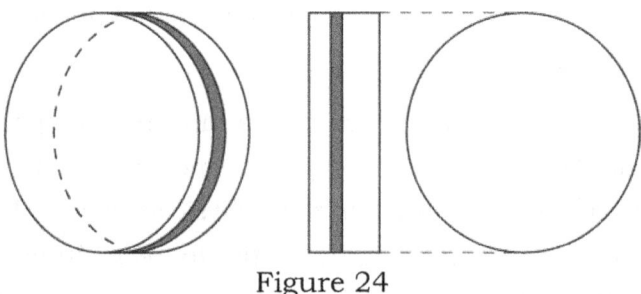

Figure 24

The neutrino c–ring merges with the intrinsic electron c-ring, providing the corset function. In parallel, the neutrino aphysical cylinder merges with the intrinsic electron aphysical cylinder also providing synchronized aphysical corset function.

Two charges, electric charge and weak electric charge, are fundamentally of the same electromagnetism nature. Electron magnetostatic field and neutrino weak magnetostatic field are coupled and merged into the single magnetostatic field.

The neutrino corset function explains the fundamental role of neutrino in particle physics. As we shall see later, the corset function is incorporated in other composite fermions: leptons and quarks.

10.3   Weak Gauge Boson W as SM Fundamental Misconception

SM muon decay diagram is presented in Figure 25. Such diagrams are common in particle physics. They are part of the electroweak unification theory; they are in published particle physics papers and in all particle physics textbooks.

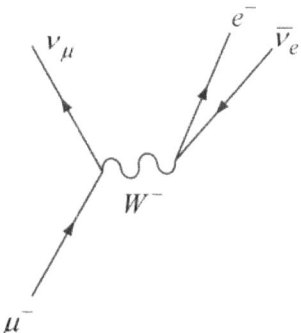

Figure 25

Muon decay and production of the electron, according to Standard Model

127

By reviewing the diagram shown in Figure 25, one can hardly find anything correct. *The diagram is SM collection of fundamental misconceptions.*

W boson, or massive weak gauge boson, carrying negative electric charge and weak isospin is an example of the glaring contradiction. By now, the reader knows that when I say a charge, I mean a fermionic inner structure. How can W weak gauge boson carry two fermionic structures, electric charge and weak isospin? This is an absurdity to the extreme.

SM also states that W boson has mass of 80.355 GeV/c². What is the origin and nature of this mass?

According to SM, W weak gauge boson mass of 80.355 GeV/c² is mass borrowed from the vacuum for the period of 10^{-24} sec and then, to be consistent with the Heisenberg uncertainty principle, promptly returned back. The problem with such borrowing and returning is violation of the energy conservation law. One cannot hide oneself behind the probabilistic uncertainty principle, which is not even applicable to individual quantum entities and individual quantum interactions.

10.4 AQM Presentation and Interpretation of the Muon Decay and the Formation of the Electron

AQM denies existence of quantum fluctuations in the vacuum, such as emerging and disappearing pairs of particle-antiparticle. In order to create a pair of particle-antiparticle, external energy and its carrier are required. The vacuum is not "replete with activities". The vacuum is quiet and majestically empty.

Virtual energy does not exist.

The vacuum does not interact with elementary particles.

The conceptual power of AQM is in its ontology and spacetime visualization. Individual elementary quantum processes can be visualized in all details in spacetime dynamics.

Here I am going to use this power for the visual presentation of the electron formation in muon decay.

The muon is a composite elementary particle consisting of bound intrinsic muon $\hat{\mu}^-$ and bound muon neutrino v_μ. The muon inner structure is identical to the electron inner structure but at higher self-energy level. The electron is the ground state lepton with self-mass of 0.511 MeV/c², while muon is a lepton at the next higher energy level with self-mass of 106 MeV/c² and an average lifetime of 2.2 microseconds. Muon decays into intrinsic muon $\hat{\mu}^-$ and muon neutrino v_μ. Consistent with AQM interpretation, the process of muon decay and the electron formation are shown in Figure 26.

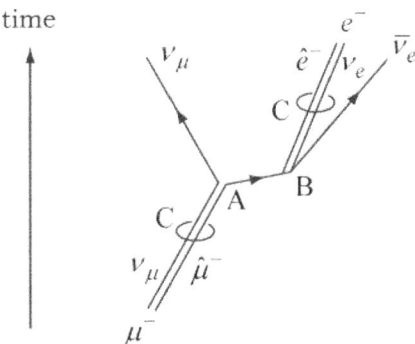

Figure 26

Muon decay and formation of electron

In AQM diagrams, the time arrow is always shown forward regardless whether we are dealing with particles or antiparticles. In this respect, a revision of a Feynman rule is overdue. Nature tells us that the direction of time is always forward. There is no reason to confuse the future generation of physicists and the general public.

Let us review and analyze muon decay and the formation of the electron as presented in diagram in Figure 26, as follows:

- Point A.

Lifetime of individual muon is determined by its initial value of individual parameter PP3 acquired at the muon formation, and can be explained by the following mechanism. Over the lifetime of an individual muon, a neutrino c-ring is drifting along muon c-ring in the direction to its nearest edge toward energetically more favorable state. There exists a force of unknown nature, perhaps elementary consciousness, which relentlessly drives a neutrino c-ring along an intrinsic muon c-ring toward its nearest edge. Once neutrino c-ring is over the edge (see point A), muon expands at nearly the speed of light within 10^{-20} seconds from its corset entrapment and reaches its classical c-ring radius of 2×1.86^{-15} meters.

Initial value of individual parameter PP3 at the muon formation defines its lifetime. There are no two muons with identical lifetime. Each muon is unique.

- Time segment AB.

Muon neutrino breaks away free carrying a portion of muon self-energy.

Intrinsic muon $\hat{\mu}^-$ transforms itself into intrinsic electron $\hat{e}^-$ by expanding radially almost with the speed of light and reaching its upper classical c-ring radius limit of $2 \times 3.86 \times 10^{-13}$ m within 1.3×10^{-21} sec, releasing some of its self-energy, forming and energizing the formation of electron neutrino-antineutrino pair.

- Point B. The formation of the electron

At point B, a pair of electron Majorana neutrinos is formed with opposite helicities (R-H, L-H) and opposite magnetic momenta. Intrinsic electron $\hat{e}^-$ selects a neutrino partner v_e with *compatible helicity*. From that moment the neutrino partner, the electron antineutrino $\overline{v}_e$ with opposite helicity, is set free. Intrinsic electron always finds a compatible neutrino partner with the same helicity and aligned magnetic moment. Original intrinsic electron goes through a chain of decays:

$$\tau^- \rightarrow \mu^- \rightarrow e^- .$$

Its helicity remains intact in a given frame of reference.

Both, the intrinsic electron and the electron neutrino, form the electron by acquiring *co-equal quantum state:* equal in Compton radius, equal in Compton angular velocity, with same helicity, equal in peripheral velocity v = c, and aligned and coupled magnetic momenta. Co-equal quantum state begins with initial classical Compton radius $2 \times 3.86 \times 10^{-13}$ meters.

- Time segment beyond B

Electron neutrino magnetostatic field fuses into intrinsic electron magnetostatic field. Then the electron neutrino proceeds with the corset action, which is a result of the small difference in values between internal and external magnetostatic fields at the c-ring surface. The corset action, with almost the speed of light, brings radius of newly formed electron down to unknown size by many orders of

131

magnitude, possibly in the range of 10^{-22} m within 1.3×10^{-21} seconds. The process of the electron formation is now complete. Simultaneously, electron antineutrino proceeds with its own additional corset action and is released free. The electron antineutrino is reduced to unknown size.

Observation of a single electron in a Penning trap suggests the upper limit of electron radius as 10^{-22} meters ([4], Dehmelt H. (1987)). Here we have another example showing that in AQM, classical-quantum divide does not exist; classical Compton c-ring radius of $2 \times 3.86 \times 10^{-13}$ meters at point A is transformed by the corset action into a quantum radius of approximately 10^{-22} meters.

Table 10-A

Comparison of the muon decay and the electron formation.

SM versus AQM

SM interpretation	AQM interpretation
1. $\mu^- \rightarrow v_\mu + W^-$ W^- is intermediate vector boson with spin $\hbar$ and electric charge e^-, weak isospin, and virtual mass 80.355 GeV/c².	1. $\mu^- \rightarrow v_\mu + \hat{\mu}^-$ $\hat{\mu}^-$ is intrinsic muon with the inner structure, spin $\hbar/2$, and self-mass of 106 MeV/c². The origin of self-mass is 100% electromagnetism.
2. $W^- \rightarrow e^- + \bar{v}_e$ Decay results in electron and electron antineutrino.	2. $\hat{\mu}^- \rightarrow \hat{e}^- + v_e + \bar{v}_e = e^- + \bar{v}_e$ $e^- = \left\{ \begin{matrix} \hat{e}^- \\ v_e \end{matrix} \right\}$

SM presents the electron as a fundamental fermion with mass 0.511 MeV/c². The origin of mass is explained by the Higgs mechanism.	Intrinsic muon releases a portion of its self-energy (self-mass), transforms itself into intrinsic electron with self-mass 0.511 MeV/c² and produces a pair of electron neutrino – electron antineutrino. Intrinsic electron acquires neutrino transforming itself into electron, a composite fermion of electromagnetism.
3. Where is electron neutrino? It is missing.	3. Electron antineutrino breaks away free carrying some self-energy.
4. The vacuum is dynamic. It is replete with activity, virtual quantum fluctuations, virtual energy, and emerging and disappearing particle-antiparticle pairs.	4. The vacuum is quiet and majestically empty. The vacuum does not interact with elementary particles. Vacuum energy density is absolute zero.

10.5 Spacetime Dynamics of Muon Decay and the Formation of the Electron

Figure 27 shows the muon decay and the formation of the electron in spacetime dynamics in 8 steps with only c-rings shown.

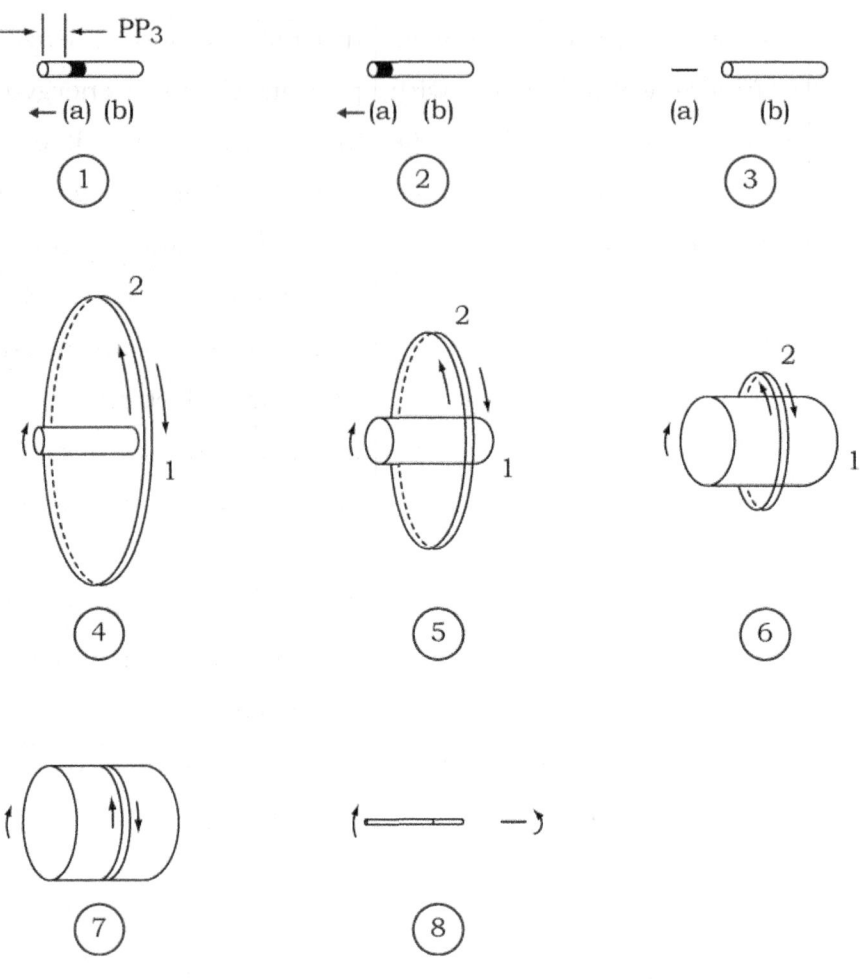

Figure 27

Explanation to Figure 27:

1. Muon μ^- is formed from a previous high energy elementary quantum process such as tau decay. The initial value of the position parameter PP3 defines the lifetime of individual muon. Muon neutrino c-ring (a) is drifting along intrinsic muon c-ring (b) toward its nearest edge in the direction of energetically more advantageous state.

2. Muon neutrino c-ring (a) reaches the edge of intrinsic muon c-ring (b) – muon life is over.

3. Muon neutrino c-ring (a) falls over the edge. It is a critical point. Muon neutrino separates from intrinsic muon and experiences additional self-corset action, additionally reducing its size. Muon is split into intrinsic muon (b) and muon neutrino (a). Intrinsic muon increases in size releasing its self-energy.

4. Intrinsic muon (1) continues release of its self-energy creating an embryo of electron neutrino – electron antineutrino pair (2). The embryo consists of four embryo constituents in a form of two embryo pairs: (I) intrinsic neutrino-intrinsic antineutrino, and (II) intrinsic neutrino-intrinsic antineutrino with reversed helicities relative to pair (I). Both pairs (I) and (II) are equally gaining self-energies. All individual intrinsic properties, such as weak electric charges, spins, and magnetic momenta, remain balanced in accordance with the conservation laws.

It is remarkable example of the origin of Majorana pair.

5. Intrinsic muon (1) continues release of its self-energy, transforming itself into intrinsic electron and transferring more self-energy to electron neutrino – antineutrino pair.

6. Process continues: intrinsic muon/electron (1) releases its self-energy and increases in size; electron neutrino-antineutrino pair is gaining self-energy. Both systems (1) and (2) are moving radially toward each other.

7. The classical Compton c-ring radius of $2 \times 3.86 \times 10^{-13}$ meters is reached and is equal for both systems. Distribution of self-mass between both systems is completed.

8. Intrinsic electron (1) selects one component (partner) of pair with the same helicity, called from that moment as electron neutrino. The second component called antineutrino is set free. After the electron is formed, it experiences a strong corset action triggered by electron neutrino, reducing its size to quantum Compton radius of approximately 10^{-22} meters. Free electron antineutrino is reduced to even smaller size by its own additional self-corset action and is separated free from the electron.

Both, electron neutrino and electron antineutrino are Majorana particles with opposite helicities. It is an example of how one inner structure can be transformed into other inner structures. No transformation of the inner structure into a field or transformation of a field into the inner structure are observed.

A question of "why classical Compton c-ring radius of $2 \times 3.86 \times 10^{-13}$ meters is selected by Nature" has the same explanation as "why is the energy of the first orbit in the hydrogen atom equal to -13.6 eV".

Above is an example of muon decay and the formation of the electron and the antineutrino:

$$\mu^- \rightarrow \hat{\mu}^- + \nu_\mu \qquad \hat{\mu}^- \rightarrow \hat{e}_e^-$$

$$\hat{e}_e^- \rightarrow \hat{e}_e^- + v_e + \overline{v}_e \rightarrow \left\{ \begin{array}{c} \hat{e}_e^- \\ v_e \end{array} \right\} + \overline{v}_e = e^- + \overline{v}_e$$

◆ ◆ ◆

Postulate 68. *The Origin of duo-fermions.*

Fermions of duo-configuration, such as duo-electron or neutrino (duo-electrino), are produced in pairs as particle and antiparticle. One cannot tell which is particle and which is antiparticle until their formation is completed, and one of them is selected based on identical helicity.

As shown in Figure 26, in the case of the formation of the electron, the selection depends on intrinsic electron helicity. One intrinsic electron selects a duo-partner with the same helicity, then the remaining duo-partner is called antiduo-particle (antineutrino, or antiduo-electrino).

An example of the origin of duo-fermions is given in the formation of the electron (see Figure 26).

As intrinsic muon (1) continues release of its self-energy creating an embryo of electron neutrino – electron antineutrino pair (2). The embryo consists of four embryo constituents in a form of two embryo pairs: (I) intrinsic neutrino-intrinsic antineutrino, and (II) intrinsic neutrino-intrinsic antineutrino with reversed helicities relative to pair (I). Both pairs (I) and (II) are equally gaining self-energies. All individual intrinsic properties, such as weak electric charges, spins, and magnetic momenta, remain balanced in accordance with the conservation laws.

137

Duo-fermion has a unique property – self-corset function. Under self-corset action duo-fermion is brought into quantum world – it acquires extraordinary small cross-section and exceptional stability. Duo-fermion has zero spin ("hidden spin"), zero electrostatic field, and double magnetostatic field producing double magnetic moment.

In particle physics existence and role of duo-electron is not established yet.

Until this study, a fundamental role of neutrino in particle physics is an open question with no answer.

♦ ♦ ♦

♦ ♦ ♦

Postulate 69. *Complete definition of the Corset: Concept, Dynamics, Mechanism, and Significance.*

Here we present a specific example of dynamics of muon decay and the formation of the electron.

- *Total spacetime dynamics is shown in Figures 26 and 27 for special case of the muon decay and the formation of the electron.*

- *Muon μ decays into intrinsic muon $\hat{\mu}^{-}$ and muon neutrino v_{μ} with initial combined self-energy of 106 MeV.*

- *Within 1.3×10^{-21} sec, intrinsic muon is transformed into intrinsic electron $\hat{e}^{-}$ and a pair of electron neutrino-electron antineutrino, v_{e}, $\bar{v}_{e}$.*

- *During the transition process, $\hat{\mu}^- \rightarrow \hat{e}^-$, the release of self-energy brings intrinsic electron c-ring and electron neutrino c-ring into co-equal quantum state: total alignment of trajectories, axes, and positions; same handedness; same peripheral velocity $v = c$; same direction of magnetic fields; same Compton radii of $2 \times 3.86 \ 10^{-13}$ meters, and same Compton angular velocities.*

- *The initial c-ring radius of $2 \times 3.86 \ 10^{-13}$ meters is a classical radius. It can be found on the basis of classical electrodynamics.*

- *Once co-equal quantum state is attained (see point 7, Figure 27), intrinsic electron magnetostatic field is fused with electron neutrino magnetostatic field.*

- *The corset action is triggered.*

- *The corset action is manifested by inward negative binding force bringing a newly formed electron down to the size of approximately 10^{-22} meters within 1.3×10^{-21} sec.*

- *Intrinsic electron/electron neutrino ratio of self-energies and ratio of spins are equal to ratio of their respective Planck constants.*

- *The corset action brings the electron from classical size down to its quantum size, with a difference of at least ten orders of magnitude.*

- *The corset action is not accompanied by energy emission or energy absorption preserving classical self-energy at quantum level.*

139

- *In addition to muon decay, the corset action plays a similar fundamental role in the formation of composite fermions, such as leptons and quarks, providing temporal or permanent stability.*

- *The corset action shows the fundamental role of the neutrino in particle physics as the triggering, binding, and stabilizing force.*

◆ ◆ ◆

10.6   The Dirac Equation and Its Enormous Unused Potential

In particle physics, the Dirac equation is a relativistic quantum wave equation derived by British physicist Paul Dirac in 1928. It is considered to be one of the greatest triumphs in theoretical physics and a fundamental contribution to quantum mechanics and quantum electrodynamics.

It is widely accepted that the equation describes the electron and, equally, the positron. The equation provides the description of the intrinsic spin and the intrinsic magnetic moment.

In some forms of the equation, an additional term is attached artificially in order to "help" the description of "the anomaly" in the electron magnetic moment.

"The anomaly" in electron magnetic moment is another SM fundamental misconception.

The Dirac equation is an important quantum probabilistic application tool in particle physics. *However, the Dirac equation's unused enormous potential has yet to be recognized.*

It would require the profound understanding of inner structures for each class of elementary particles and the functional relationship among constituents. *Such understanding comes from Aphysical Quantum Mechanics.*

In fact, the Dirac equation in its present form does not describe the electron, which is a big surprise to theorists. It describes the intrinsic electron, a constituent of the electron.

However, the Dirac equation is adaptable for the description of all intrinsic fermions (and intrinsic antifermions) for any of three fundamental forces they represent:

- for electromagnetism: Planck constant $\hbar$, electric charge $-e$ and intrinsic electron self-mass (or other self-masses related to intrinsic leptons);

- for weak electromagnetism: weak Planck constant h_w, weak electric charge $-\beta e$, and intrinsic neutrino self-mass (*which was derived in AQM with great precision*);

- for strong force: strong Planck constant h_s, color charge and self-mass for intrinsic colotron (yet to be discovered).

The SM statement that all fermions regardless of which fundamental force they represent have the same spin $-\frac{1}{2}\hbar$ is another SM fundamental misconception.

◆ ◆ ◆

Postulate 70. *Expanding Dirac equation to all classes of elementary particles of matter.*

- *The Dirac equation does not describe the electron – it describes the intrinsic electron.*
- *The profound understanding of inner structures and functional relationship among constituents for each class of elementary particles is a key for the expansion of the Dirac equation to its full potential. Such understanding comes from Aphysical Quantum Mechanics.*
- *In the final analysis, the inner structure of the composite elementary particle consists of intrinsic fermions such as intrinsic electron and intrinsic neutrino, intrinsic fermions with fractional charges, and intrinsic fermions of strong force.*
- *Intrinsic fermions are the simplest form of matter, the final frontier on the way toward the smallest and the simplest things.*
- *Each type of intrinsic fermion of a given fundamental force is described by its Planck constant, elementary charge of the force it represents, and by its self-mass derived from experiment.*
- *The Dirac equation is directly applicable to any fermion, intrinsic or composite, regardless of the fundamental force it represents.*
- *Based on the principle of linearity, one can expand application of the Dirac equation to any composite fermion. If the Dirac equation can describe any intrinsic fermion constituent sepa-*

rately, it also can describe their linear combination in a form of composite fermions.

- *All of the above is symmetrically applicable to antifermions.*

◆ ◆ ◆

It is a challenging task for undergraduates and graduates in their desire to contribute to fundamental physics to expand the Dirac equation to its full potential without waiting for theorists who have found themselves in the Never-Never Land.

10.7   The Anomalous Magnetic Moment of the Electron as SM Misconception

Dirac theory predict that the electron magnetic moment is defined by the relationship

$$\bar{\mu} = g \frac{Q}{2m} \bar{S} \,,$$

where g = 2 (g – factor), but experimentally is known to be greater than 2, expressed as

$$\frac{g-2}{2} = \alpha \,.$$

This small value α is *the anomaly.*

According to SM, the anomaly of the electron magnetic moment arises from infinite number of QED quantum contributions in the dynamic vacuum described by Feynman diagrams.

For several decades, quantum theorists have undertaken a titanic effort to calculate the anomaly with ever higher accuracy using thousands of Feynman high level loop diagrams, hoping to reach "new physics". No new physics has ever been found. Loop diagrams are absurdities. For example, it requires an immediate photon emitted by electron to turn around and be absorbed by the same electron.

Surprisingly, the result of these calculations is consistent (to eleven decimal places) with high precision experimental results.

According to AQM, there are no quantum fluctuations in the vacuum. The quantum vacuum is majestically empty and quiet. It is up to quantum theorists to explain how they have succeeded to get such a remarkable agreement with the experiment. Such calculations are scientifically "illegitimate". The vacuum does not have QED quantum fluctuations.

Unfortunately, such titanic effort has no value to future generations of physicists. SM "amazing" triumph is not really a triumph.

The so-called "anomaly" of the electron magnetic moment is the magnetic moment of the electron neutrino, a constituent of the electron.

Those who disagree and insist that the neutrino magnetic moment can be calculated on a basis of QED quantum contributions, must be ready to affirm that similar calculations can be done to determine the intrinsic electron magnetic moment.

10.8   AQM Arrives at Super Accurate Values of Electron Neutrino Self-Mass, Weak Electric Charge, and Weak Planck Constant

My task is to arrive at values of the most important properties of the electron neutrino such as

- Weak electric charge w in terms of fractional electric charge e;

- Planck constant for weak electromagnetism $\hbar_w$;

- Electron neutrino self-mass ; and

- Electron neutrino magnetic moment M.

I performed this important extraordinary task and achieved extreme accuracy using college mathematics (see Postulate 72). No high order of mathematical fog is acceptable, although my task needs profound understanding of inner structures and super intuition.

This is accomplished on the basis of available experimental data and *profound* understanding of the electron inner structure, such as

- Experimental data of electron self-mass

$$m_e = 0.510\ 998\ 910\ eV$$

- Experimental data for electron constant α_e

$$\alpha_e = 0.001\ 159\ 652\ 188\ ,$$

obtained experimentally by Dehmelt group in 1987 using a Penning trap [5].

- Profound understanding of weak fundamental force as a branch of electromagnetism coined by me as *weak electromagnetism;* and

- Profound understanding of the electron inner structure, both structural and functional.

For many decades particle physicists have been searching for the electron neutrino mass. According to SM, the electron neutrino has zero mass, which is fundamental misconception. Extensive experimental and theoretical effort leads to estimating neutrino mass in a range of 0.2 – 2.0 eV.

I have no choice but state that extensive theoretical and experimental work on "the neutrino oscillations" has introduced a great degree of additional confusion into the neutrino mass problem. The neutrino oscillations theory is not valid scientifically. Individual neutrinos traveling free in space do not change their self-mass unless they have sufficient energy for formation of other inner structures. The process is unlikely and, in any case, is not oscillatory.

In stark contrast, the measurements of the magnetic moment of the electron using a Penning trap, is a great triumph of experimental particle physics.

A single electron can be trapped indefinitely in the combination of a homogeneous magnetic field and an electrostatic quadrupole potential, known as a Penning trap.

It is an example of an experiment with individual electrons. An electron is real before measurement, during measurement and after measurement. It exists even without being measured, the thought, I am confident, is upsetting to quantum positivists.

It is not a statistical electron burdened by wave function. It is an individual electron bound in a Penning trap where electron properties are measured with extraordinary precision to eleven digital places.

10.9   SM History of Anomalous Magnetic Moment of the Electron

According to SM, the electron is the fundamental fermion of electromagnetism. It is point-like and has no inner structure. Magnetic moment of the electron is equal to one Bohr magneton.

However, precise experimental measurements of magnetic moment of a free electron performed by Dehmelt group in 1987 using a Penning trap [5] shows that free electron has anomaly

$$\alpha_e = 0.001\ 159\ 652\ 188 \qquad (1)$$

This result (1) stood for 24 years until it was modestly improved experimentally by Gabrielse group in 2011 [6]

$$\alpha_e = 1.001\ 159\ 652\ 180\ 73(28) \qquad (2)$$

The result (2) remains the most recent experimental result. However, there was collaboration between experimental Gabrielse group and theoretical Kinoshita group [7]. In the situation when theoretical result cannot be independently reproduced, it is problematic. Therefore, my reference is and always will be Dehmelt group result (1).

This was a triumph of experimental physics. Now it was up to theorists explain the anomaly. There is a long history of such theoretical explanations. Anomalous magnetic moment is explained on a basis of virtual quantum fluctuations interacting with electron and contributing to its magnetic moment.

It is stated that in case of electron it is sufficient to consider only QED contributions from virtual pairs of electron-positron and virtual photons. Such contributions to electron magnetic moment are calculated using Feynman diagrams.

According to Feynman's rule of calculation, to obtain the most accurate result, one has to integrate all values of QED contributions, including high order loops, over the entire range of energies. It is a mind-boggling task (Thank you, Prof. Feynman for your unrestrained imagination!).

Several groups of theoretical physicists have been actively involved in such calculations. The most notable of them is the group of Tatsumi Aoyama, Masashi Hayakawa, Toichiro Kinoshita, and Makiko Nio (2012). "Tenth-Order QED Contribution to the Electron g-2 and an Improved Value of the Fine Structure Constant", *Physical Review Letters.* **109** (11): 111807 [8].

To complete their calculations, the Kinoshita group spent more than 20 years using a total of 16,676 Feynman diagrams including tenth-order QED loop contributions. This was an incredible time consuming effort. In 2012, Kinoshita group published their best computed value of the anomaly [8]:

$$\alpha_e = 0.001\ 159\ 652\ 181\ 13(86). \qquad (3)$$

This is called a triumph of QED? Not so fast!

Here is my observation and conclusion:

1. I do not doubt integrity and honesty of the authors [8] published 2012.

2. The experimental result (1) by Dehmelt group has been known since 1987. It has been the guiding star.

3. The calculation of high order QED contribution of such complexity does not have a unique pathway toward the result. In fact, such calculation has numerous pathways toward the result under the direction of the guiding star.

4. "Just because the results happen to be in agreement with experiment does not prove that one's theory is correct", stated Paul Dirac in his last paper, in 1984, on the subject of QED contributions to anomalous electron magnetic moment.

5. Those who have dedicated many years in search of new physics by calculating QED contributions, in fact, have been creating obstacles on the way to new physics without realizing it.

The best theoretical result (3) should be considered "a theoretical experiment". It can never be repeated or reproduced. Scientifically, it is not acceptable situation in the same way as any experiment which cannot be confirmed independently by others. The result (3) is scientifically illegitimate.

What can we say about *muon anomalous magnetic moment*? Here are experimental [9] and theoretical [10] results:

$$\alpha_\mu = 0.001\ 165\ 920\ 9\ (experiment) \qquad (4)$$

and

$$\alpha_\mu = 0.001\ 165\ 918\ 04\ (theory) \qquad (5)$$

The experimental result [9] was obtained in the Brookhaven National Lab. It is a challenging experiment as muon has lifetime of 2.2 μs and is 206 times heavier than electron.

Let us compare muon anomaly (4) with electron anomaly (1):

$\alpha_\mu =$	0.001 165 920 9	(experiment)	(4)
$\alpha_e =$	0.001 159 652 188	(experiment)	(1)

$$\alpha_\mu - \alpha_e = 0.000\ 006\ 268\ 7 \qquad (6)$$

I declare here that the difference (6) in experimental results is a systematic error in the BNL experiment.

According to my theorem (see Postulate 73 in Chapter 11), value of α is a constant and identical for all leptons:

$$\alpha_e = \alpha_\mu = \alpha_\tau = \alpha$$

It is explained by the fact that self-mass distribution in lepton pre-formation stage is absolutely the same for all leptons due to absolute equality in its classical Compton angular velocities of intrinsic lepton and lepton neutrino:

$$\omega_c(\hat{l}) = \omega_c(\nu_l)$$

What is curious, or maybe not, is that muon calculation (5) faithfully reproduces the BNL systematic experimental error.

However, if a shift in value of α actually exists along the lepton chain, such as $\alpha_e < \alpha_\mu < \alpha_\tau$, then Nature's design of leptons is not perfect. Is it possible? Very unlikely.

The experimental result (1) which has been shining upon us since 1987, combined with AQM understanding of inner structures of intrinsic leptons and lepton neutrinos, and understanding of the process of lepton formation, allows us derive such fundamental values as weak charge, weak Planck constant, and lepton neutrinos self-mass, with high precision.

The experimental result (1) obtained by Dehmelt group in 1987 must be written in gold.

Without experimental data, progress in theoretical physics is impossible. Eventually it would lead to imaginary worlds and pseudo-realities, as is already happening.

As in case of the aphysical quantum optics experiment (see Volume 2) [12], it is an example of an experiment with individual elementary particles and individual interactions. *Potential for such experiments is enormous.*

A surprise comes from AQM, a deeper quantum theory. The AQM analysis is based on profound understanding of the electron inner structure.

The AQM electron is a composite fermion of electromagnetism. Its inner structure has two constituents: the intrinsic electron and the electron neutrino. The electron neutrino has duo configuration consisting of the intrinsic electron neutrino and the intrinsic electron antineutrino with opposite helicity.

A free electron traveling in space has the complete definition in the quantum reality, including well-defined position with zero momentum in the frame of self-reference. A free full-fledged electron travels in free space along its unique trajectory. The vacuum is empty. Electron is not surrounded by clouds of virtual particles. There are no virtual quantum fluctuations. One does not need Feynman diagrams to calculate the so-called "anomaly". What is called "anomaly" is in fact the magnetic moment of electron neutrino.

10.10   Electron Diffraction and Interference

The mechanism of electron diffraction/interference is principally different compared with the photon. In case of photons, the aphysical energy cylinder plays the principal role in interactions with aphysical periodic structures, such as multi-slit or crystal of comparable periodicity relative to the aphysical cylinder circumference.

In case of electrons, the aphysical cylinder plays no role in diffraction/interference. Its quantum radius is 10^{-22} meters at upper limit which is six or more orders of magnitude smaller than, for example, the size of proton. Such periodic aphysical structures do not exist.

The key player for electron diffraction/interference is *the kinetic energy aphysical column*. The length of the aphysical energy column is defined by the de Broglie relation

$$L_{ka} = \lambda = \frac{h}{|p|}$$

In the electron, the intrinsic physical energy E_{sp} and the kinetic physical energy E_{kp} are contained in the electron c-ring with total physical energy E_p equal, $E_p = E_{sp} + E_{kp}$.

On the other hand, electron intrinsic aphysical spin energy $E_{sa} = E_{sp}/U$ is contained in the aphysical energy cylinder separately from the aphysical kinetic energy $E_{ka} = E_{kp}/U$ which is contained in the aphysical kinetic energy column with total electron aphysical energy E_a equal

$$E_a = E_{sa} + E_{ka} = \frac{E_p}{U},$$

where U is the universal constant (see Figure 28).

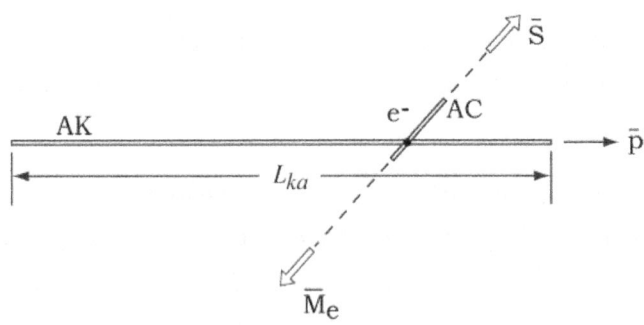

Figure 28
AC-electron aphysical cylinder;
AK – electron kinetic energy column

152

Historically, first electron diffraction experiment was performed by Davisson and Germer in 1927, thus confirming the de Broglie hypothesis.

Since then numerous electron diffraction experiments have been performed both, in transmission and backscatter modes.

In case of the transmission mode, as electron passes through a thin layer of crystalline material, a diffraction pattern is observed on a detection screen.

According to AQM, the electron diffraction pattern is a result of aphysical-aphysical energy interaction of self-entangled electron with a periodic crystalline structure. In the Davisson-Germer experiment, the spacing of the nickel crystalline planes is equal to d=0.091 nm, and electron kinetic aphysical length (or de Broglie wavelength) L_{ka} is equal to 0.167 nm for electron kinetic energy of 54 eV.

In cases of $d \gg L_{ka}$ or $d \ll L_{ka}$, no electron diffraction/interference is possible. Periodicity of the crystalline structure and the electron aphysical length must be comparable.

In contrast to optical materials, the crystalline nickel has a limited electron transparency. Therefore, the nickel sample must be thin, on the order of 100 nm. Electron diffraction involves all three types of interactions:

- physical-physical;
- aphysical-physical;
- aphysical-aphysical.

In quantum optics (or optics in general), only aphysical-aphysical interaction is present. That is why I call quantum optics *aphysical quantum optics*. Only aphysical-aphysical interaction contributes to electron diffraction/interference. In such case, electron is self-entangled traveling along countless trajectories.

If physical-physical interaction of the host with the crystalline physical structure occurs, an electron and all its a-fractions are knocked-out of diffraction process. That is why a crystalline sample must be thin. Aphysical-physical interaction of electron a-fraction with crystalline physical structure is not permitted. Such a-fraction is instantaneously recalled by the host and is not available to contribute to aphysical diffraction.

Let us consider a diffraction experiment with the ideal crystal of nickel in the transmission mode based on AQM interpretation. For simplicity, let us take Davisson and Germer arrangements with the spacing of the nickel crystalline plane d = 0.1 nm and de Broglie wavelength L_{ka} = 0.167 nm for the electron kinetic energy of 54 eV.

We send a single electron to nickel crystal. As it crosses the first crystalline plane, electron becomes self-entangled. After crossing each subsequent crystalline plane, the number of a-fractions doubles producing a sort of chain-reaction. In the case of the perfect crystal, depending on crystal thickness t_c, one will have following number of aphysical a-fractions N at the crystal output:

$$t_c = 1 \text{ nm} \qquad N = 10^3;$$
$$t_c = 2 \text{ nm} \qquad N = 10^6;$$
$$t_c = 3 \text{ nm} \qquad N = 10^9;$$
$$t_c = 4 \text{ nm} \qquad N = 10^{12}.$$

$$t_c = 100 \text{ nm} \qquad N = 10^{300}$$

If we had the aphysical detectors, we would be able to observe *the entire diffraction pattern produced by a single electron in a nickel crystal with thickness of 5 nm.* However, such diffraction pattern would be instantaneously cancelled at the moment electron c-ring (the host) strikes the detection screen. This is not a big challenge – the c-ring could be "gently" removed from space between the crystal and the detection screen using magnetic field. *The real challenge for the future experimental physicists is to develop the aphysical detection technique.*

In a real crystal one should expect substational losses of a-fractions and hosts.

10.11 Fundamental Re-Interpretation of Louis de Broglie Relation

According to the de Broglie relation, a particle with mass m and momentum $\bar{p}$ has a wavelength λ:

$$\lambda = \frac{h}{|\bar{p}|}$$

In AQM, this relation acquires new fundamental meaning and significance.

A particle with self-mass m, momentum $\bar{p}$, and physical kinetic energy E_{kp} has a column of aphysical kinetic energy $E_{ka} = E_{kp}/U$ projected in the direction of the momentum vector.

For example, an electron with physical kinetic energy $E_{kp} = 13.6$ eV has aphysical kinetic energy column length L_{ka} equal to the circumference of the first electron orbit of the hydrogen atom ($L_{ka} = 0.33$ nm). It is an example of how aphysical energy imitates angular momentum of physical energy.

155

♦ ♦ ♦

Postulate 71. *Definition of Aphysical Kinetic Energy Column and Properties.*

Consider the case of the full-fledged truly elementary particle (intrinsic fermion), traveling with momentum p and relativistic velocity v, where c – v << c. In such relativistic case, the length of the physical energy c-ring experiences relativistic contraction, while the length of aphysical energy cylinder remains unchanged because aphysical energy cylinder has no relativistic properties. Instead, the aphysical kinetic energy column is formed and projected in the direction of momentum p with the length derived from the de Broglie relation, $L_{ka} = \dfrac{h}{|p|}$.

The physical energy c-ring carries both physical spin energy and physical kinetic energy.

Aphysical energy is space-separated in two entities:
- *aphysical energy cylinder carrying aphysical spin energy; and*
- *aphysical kinetic energy column carrying aphysical kinetic energy.*

Physical spin energy and aphysical spin energy are not subject to relativistic influence, which means that linear velocity has no influence on values of spin or magnetic moment.

Physical kinetic energy in the c-ring and aphysical energy column have relativistic contraction (see Figure 29 (a, b, c)).

156

In contrast to QM with its numerous enigmas and various in-terpretations, Einstein's special relativity has so far remained intact and uncontroversial. However, special relativity is about to be sub-jected to AQM fundamental test.

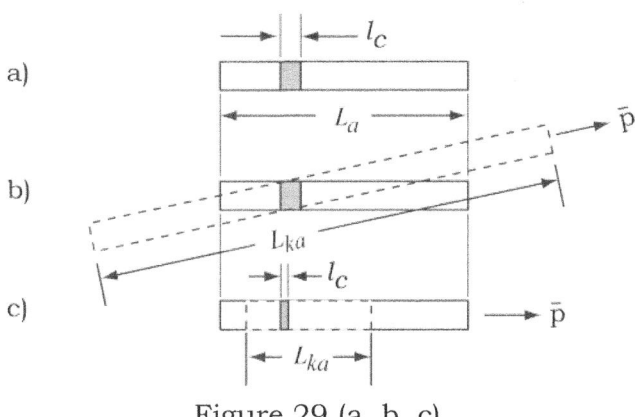

Figure 29 (a, b, c).

Let me further explain the statements presented in Postulate 71 and shown in Figures 29 (a, b, c).

There are three cases: (a) stationary, (b) non-relativistic, and (c) relativistic.

A. Stationary case with v = 0, see Figure 29 (a).

The particle has physical spin energy E_{sp} and aphysical spin energy $E_{sa} = E_{sp}/U$. The length of physical spinning c-ring is shown as l_c and the length of aphysical spinning cylinder is shown as L_a.

B. Non-relativistic case with velocity v << c and momentum $\bar{p}$, see Figure 29 (b).

The c-ring carries both physical spin energy E_{sp} and physical kinetic energy E_{kp}. Total physical energy E_p is equal to sum of physical spin energy E_{sp} and physical kinetic energy E_{kp}:

$$E_p = E_{sp} + E_{kp}$$

C. Relativistic case with velocity c-v<<c and momentum p.

Both the c-ring and aphysical kinetic energy column experience relativistic contraction. The aphysical spin energy cylinder remains unchanged.

10.12 The Electron-Positron Collision and the Transformation of Electromagnetic Energy

Here I describe and explain properties of the photon physical c-ring separately from properties of the photon aphysical cylinder (AC) and performance of the elementary consciousness. To be consistent with special relativity, the photon physical energy c-ring has zero length in the direction of travel. In fact, there is no mathematical complexity here. Furthermore, the photon c-ring, as any bosonic c-ring, has zero cross-section. All photon physical energy is confined within the photon c-ring with zero cross-section and, hence, zero space.

The photon c-ring is a bundle of physical energy spinning with the circumferential speed of light at the Compton angular velocity $\omega = c / r$, where r is the Compton radius of the bundle. To satisfy special relativity, the bundle must have zero cross-section.

158

According to SM, the photon is called the quantum of electromagnetic energy (or quantum of electromagnetic fields) and is the force carrier for electromagnetism. Something is wrong with the SM photon definition. It is another SM fundamental misconception. The properties of electromagnetic energy would not allow such energy to be totally "squeezed" into the c-ring or, in other words, to be totally confined in zero space.

According to AQM, the photon is the quantum of inverted electromagnetic energy.

Here I introduce concepts of inversion and the inverted electromagnetic field (energy). One should not confuse electromagnetic energy with photon inverted electromagnetic energy. *The photon is not quantum of electromagnetic energy.*

In order to explain the fundamental nature of the inversion of the electromagnetic field (energy) into the inverted electromagnetic field (energy) confined in the c-ring with zero cross-section, let me show three examples.

Example 1. The annihilation of an electron-positron pair and production of a pair of gamma-photons (γ_e):

$$e^- + e^+ \rightarrow \gamma_e + \gamma_e \qquad (10\text{-}1)$$

The process of the annihilation, as shown in (10-1), is in agreement with SM. However, as explained later, the process is much more complex than it is assumed in SM.

As presented in AQM, the electron (or the positron) is a composite elementary particle, consisting of intrinsic electron $\hat{e}^-$ (or intrinsic positron $\hat{e}^+$) and electron neutrino ν_e (or electron antineutrino $\overline{\nu}_e$):

$$e^- = \left\{ \begin{matrix} \hat{e}^- \\ v_e \end{matrix} \right\} \quad and \quad e^+ = \left\{ \begin{matrix} \hat{e}^+ \\ \bar{v}_e \end{matrix} \right\} \qquad (10\text{-}2)$$

Both electron and positron have equal self-masses and therefore, equal radii independent of system of reference.

The process of the annihilation can be presented in detailed spacetime dynamics. In AQM, the process is not called "the annihilation" but "the transformation of electron-positron pair into pair of γ_e photons".

For the purpose of example 1, I utilized only a portion of the process, namely the physical process of the collision of intrinsic electron c-ring with intrinsic positron c-ring and the transformation into a pair of photon c-rings:

$$(\text{c-ring})_{\hat{e}^-} + (\text{c-ring})_{\hat{e}^+} \rightarrow (\text{c-ring})_{\gamma_e} + (\text{c-ring})_{\gamma_e}$$

Figure 30 shows two c-rings: $(\text{c-ring})_{\hat{e}^-}$ and $(\text{c-ring})_{\hat{e}^+}$ prior the collision heading toward each other. Each c-ring carries electrostatic energy E_e and magnetostatic energy E_m within unrestricted (unlimited) space. Both energies, E_e and E_m, are equal, $E_e = E_m$.

For each element in electrostatic and magnetostatic field configurations of the intrinsic electron c-ring there is a corresponding anti-element in electrostatic and magnetostatic field configurations of the intrinsic positron c-ring (see Figure 30).

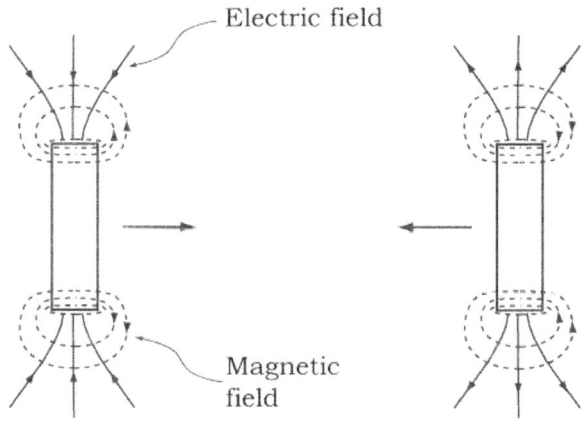

Electric field

Magnetic field

Figure 30.

Collision of intrinsic electron and intrinsic positron

(only physical energy c-rings are shown)

During the process of the transformation, intrinsic electron (and symmetrically, intrinsic positron) electrostatic and magnetostatic energies E_e and E_m are transformed into the inverted electrostatic and inverted magnetostatic energies I_e and I_m of photon c-ring, with the conservation energy law observed (see Figure 31)

$$E_e = I_e \text{ and } E_m = I_m$$

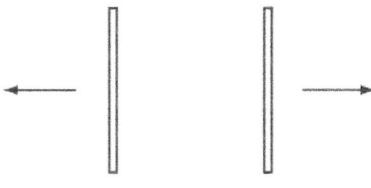

Figure 31.

Creation of γ_e photon pair from the annihilation of

electron-positron pair

(only photon c-rings are shown)

◆ ◆ ◆

Postulate 72. *AQM complete definition of the electron*

- *The electron is a composite fermion consisting of the intrinsic electron and the electron neutrino. The electron neutrino has the duo configuration. It is a duo-fermion consisting of the intrinsic electron neutrino and the intrinsic electron antineutrino with the reverse helicity and aligned magnetic moment.*

- *The components of the electron inner structure are the intrinsic electron inner structure, consisting of the c-ring and the aphysical cylinder, and the electron neutrino inner structure consisting of the c-ring and the aphysical cylinder.*

- *The electron self-energy E is equal to intrinsic electron electrostatic energy E_E and magnetostatic energy E_H, where $E_E = E_H$, and neutrino magnetostatic energy E_ν only.*

- *The electron neutrino has zero spin, zero weak electrostatic field and zero weak electric charge.*

- *The electron neutrino has the magnetic moment equal two times of weak Bohr magneton,*

$$M_\nu = \frac{\beta^2 e\hbar}{m_\nu}$$

- *The electron has self-mass of 0.511 MeV/c². AQM predicts that the neutrino contribution to electron self-mass is equal to 296 eV/ c² calculated as*

162

$$m_{\hat{e}} = \frac{m_e}{1+\beta} \qquad \text{and} \qquad \beta = \tfrac{1}{2}\,\alpha,$$

where $\alpha = 0.001\ 159\ 652\ 188$ *is measured experimentally with accuracy to ten digits in a Penning trap.*

- *The electron quantum values are:*

negative electric charge $-e$ *and zero weak electric charge* $(w - w)$, *where*

$$w = \beta e;$$

spin $S = \hbar/2$; *and magnetic moment* M_e *equal to magnetic moment of the intrinsic electron and magnetic moment of the electron neutrino,*

$$M_e = M_{\hat{e}}(1+\alpha).$$

- *At the pre-Corset action, the intrinsic electron c-ring can be described by classical electrodynamics and the neutrino c-ring can be described by weak electrodynamics.*

During electron formation at the pre-Corset stage, the electron has classical Compton radius of $2\times3.86\times10^{-13}$ *meters. At the completion of the Corset action, the electron classical radius is reduced to quantum Compton radius by at least nine orders of magnitude to upper limit of* 10^{-22} *meters, as shown in Penning trap experiments.*

- *The electron is the quantum ground state lepton.*

- *The electron has two quantum states: the full-fledged quantum state and the quantum state of self-entanglement. Diffraction and interference are explained by aphysical-aphysical interaction of self-entangled electron with the periodic structure where the kinetic energy column plays the principal role. Both, the length of the kinetic energy column and the periodicity of the periodic structure, must be comparable.*

- *The length of the kinetic energy column is defined by the de Broglie relation.*

  In the ultra-relativistic case (c-v<<c), the c-ring and the kinetic energy column experience relativistic contraction. The aphysical cylinder experiences no relativistic contraction. In the ultra-relativistic case, the c-ring, the kinetic aphysical energy column, and the aphysical cylinder are aligned in the direction of electron momentum.

- *The electron has four fundamental geometrical constants: $l_{c\hat{e}}$, l_{cv} , $L_{a\hat{e}}$, and L_{av} , each corresponding to lengths of c-rings and aphysical cylinders.*

- *The electron has three Position Parameters. No two electrons are identical – each has a set of unique Position Parameters.*

- *The orbital electron does not radiate (as explained in Volume One).*

- *The electron has in its system of self-reference (v=0) only spin energies: c-ring physical spin energy and aphysical cylinder aphysical spin energy.*

  Outside of the system of self-reference (v ≠ 0), electron acquires both physical and aphysical kinetic energies.

- *Aphysical spin and aphysical kinetic energies are space-separated between the aphysical cylinder and the aphysical kinetic column.*

◆ ◆ ◆

Chapter 11
AQM Leptons

11.1 Muon and Tau are Electrons at Two Energy Levels Up

To call leptons "flavors" would be like calling electrons on first three orbital levels in the hydrogen atom "flavors". The term "flavors" has to go.

In a given pathway of lepton decays and subsequent transformations from tau to electron, an intrinsic electron emerging from tau decay and by partially releasing self-energy converts itself into muon, then emerging from muon decay and partially releasing self-energy converts itself into electron, thus reaching its final quantum ground state. In this chain of decays and transformations, it is the same intrinsic electron with the same helicity going from one transformation state to the next, and in the process partially releasing its self-energy along the leptonic pathway.

11.2 Intrinsic Leptons, Lepton Neutrinos, and Leptons

According to SM there are three families of leptons:

     I.   electron e^- and electron neutrino ν_e;

     II.   muon μ^- and muon neutrino ν_μ;

     III.   tau τ^- and tau neutrino ν_τ.

All leptons are considered to be truly elementary particles, point-like and without inner structures. It is another SM fundamental misconception.

According to AQM, *free neutrinos* ν_e, ν_μ, ν_τ *are not leptons*. They are duo-neutrinos (Majorana particles). Each type of neutrino is a composite fermion of weak electromagnetism consisting of the intrinsic neutrino and the intrinsic antineutrino with opposite helicity. The intrinsic neutrino is the exclusive carrier of one elementary unit of negative weak electric charge (-w), like the intrinsic antineutrino is the exclusive carrier of one elementary unit of positive weak electric charge (+w).

Here is a definition of AQM leptons:

Electron e^-, muon μ^-, and tau τ^- are *composite* elementary particles, each consisting of two constituents - the intrinsic lepton and the bound lepton neutrino:

$$e^- = \left\{ \begin{array}{c} \hat{e}^- \\ \nu_e \end{array} \right\}, \qquad \mu^- = \left\{ \begin{array}{c} \hat{\mu}^- \\ \nu_\mu \end{array} \right\}, \qquad \tau^- = \left\{ \begin{array}{c} \hat{\tau}^- \\ \nu_\tau \end{array} \right\},$$

where $\hat{e}^-$, $\hat{\mu}^-$ and $\hat{\tau}^-$ are intrinsic electron, intrinsic muon, and intrinsic tau.

Intrinsic leptons are truly elementary particles, fundamental fermions of electromagnetism and exclusive carriers of one elementary unit of negative electric charge (-e), like intrinsic antileptons are fundamental fermions of electromagnetism and exclusive carriers of one elementary unit of positive electric charge (+e).

Without addition of neutrinos to their inner structures, intrinsic leptons are unstable, restless, striving to release their self-energy and transform themselves into other inner structures. With addition of neutrinos, they are transformed into leptons. The lepton neutrino performs a "corset" function bringing intrinsic lepton classical Compton radius down to quantum Compton radius, thus providing relative stability to leptons with the following experimentally established average lifetimes:

$$t_\tau = 2.9 \times 10^{-13} \text{ seconds}$$

$$t_\mu = 2.2 \times 10^{-6} \text{ seconds}$$

$$t_e = \infty$$

According to AQM, bound intrinsic leptons, constituents of leptons, have the same properties as *the generalized intrinsic electron* $\hat{e}^-$ *(g)*. Intrinsic leptons, $\hat{e}^-$, $\hat{\mu}^-$ and $\hat{\tau}^-$ have identical inner structures, each one at next levels of self-energy. Intrinsic leptons have equal length of the c-ring l_c and equal length of the aphysical cylinder L_a, as shown in Figure 32.

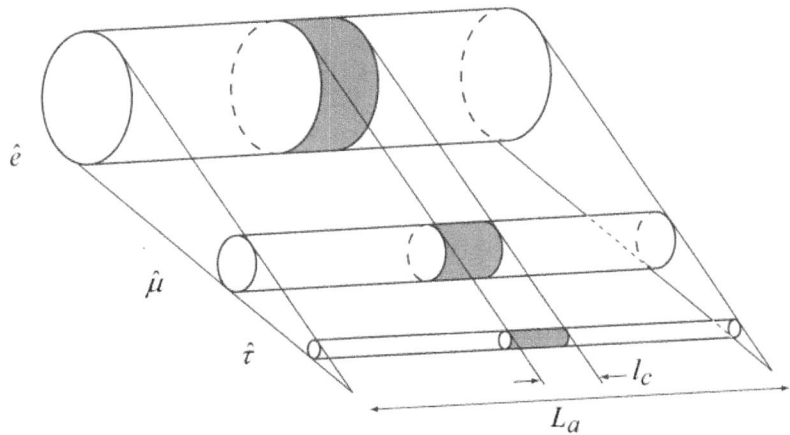

Figure 32

Inner structures of intrinsic leptons

Comment: in reality, positions of c-rings relative to aphysical cylinders are not aligned – they are irreducibly random.

The leptons are the same particle, "the electron", at three different self-energy levels:

1. Electron at 0.511 MeV;

2. Muon at 106 MeV; and

3. Tau at 1.778 GeV.

In the atom, one does not call orbiting electrons by different names depending on their orbital energy levels. We would have run out of names quickly.

11.3   The Inner Structure of Leptons

Lepton inner structures at pre-Corset state are shown in Figure 33. Pre-Corset Compton radius of each type of lepton is classical and can be calculated on basis of classical electrodynamics:

$$r_\tau = 2 \times 1.1 \times 10^{-16} \text{ m};$$

$$r_\mu = 2 \times 1.86 \times 10^{-15} \text{ m};$$

$$r_e = 2 \times 3.86 \times 10^{-13} \text{ m}.$$

The Corset action reduces classical Compton radii of leptons by many orders of magnitude to level below of 10^{-22} meters as a result of a negative binding force generated by lepton neutrino corset action. In spite of such dramatic size reduction to their quantum Compton radii at nearly the speed of light, leptons do not radiate or absorb energy. Leptons self-energy acquired at the classical energy level is preserved at quantum Compton radius.

The lepton is a composite elementary particle consisting of the intrinsic lepton and the bound lepton neutrino. That is the definition of the lepton. The intrinsic lepton inner structure consists of the physical c-ring and the aphysical cylinder. The bound lepton neutrino inner structure consists of the neutrino physical c-ring and the neutrino aphysical cylinder. The bound lepton neutrino is a composite fermion, consisting of the intrinsic neutrino and the intrinsic antineutrino with opposite helicity. Lepton neutrinos have duo configuration. They are duo-particles.

In a relativistic case, in addition to its spin energy, the c-ring acquires physical kinetic energy. The c-ring length experiences relativistic contraction while aphysical cylinder length remains constant - *it is not subject to special relativity.*

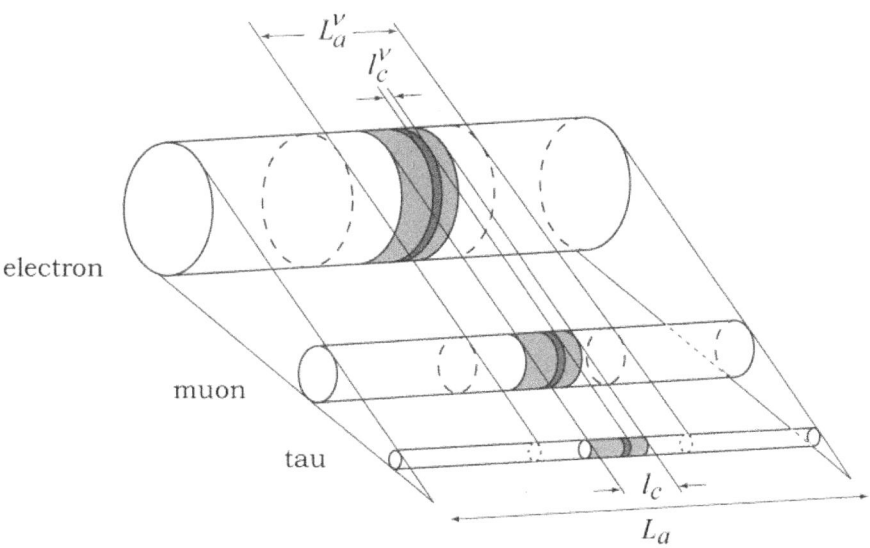

Figure 33

Inner structures of leptons

Comment: in reality, positions of c-rings relative to aphysical cylinders and c-rings relative to each other are not aligned – they are irreducibly random.

The aphysical cylinder contains only the aphysical spin energy. The aphysical kinetic energy is expressed in a form of the aphysical kinetic column with the length defined by the de Broglie relation. The aphysical spin energy and the aphysical kinetic energy are contained in two separate sub-structures - the aphysical cylinder and the aphysical kinetic column. The length of the aphysical kinetic energy column experiences relativistic contraction (see Figure 29a, b).

In super-relativistic case, c-v << c, the physical c-ring, the aphysical cylinder, and the aphysical kinetic energy column are aligned in the direction of momentum (see Figure 29 c).

11.4 The Lepton Transformation Along the Leptonic Pathway

As a deeper quantum theory, AQM is in the position to describe instant by instant spacetime dynamics for the lepton transformation along a pre-selected leptonic pathway, including transitory quantum processes occurring between stages of relative stability.

Let us consider the transformation of a high energy generalized intrinsic electron $\hat{e}^-$ (g) entering the leptonic pathway:

$$\hat{e}^-\ (g) \to \tau^- \to \mu^- \to e^- \quad (see\ Figure\ 34)$$

Of course, there are other leptonic decay modes such as tau decay into hadronic quantum states (65%), or tau decay directly into electron bypassing muon (17.85%).

The leptonic transformation process begins with a generalized intrinsic electron $\hat{e}^-$ (g) entering the pathway with self-energy much higher than tau self-energy, making a temporary stop at tau stage of relative stability and transforming itself into tau with self-energy 1.777 GeV for period of 2.9×10^{-13} sec.

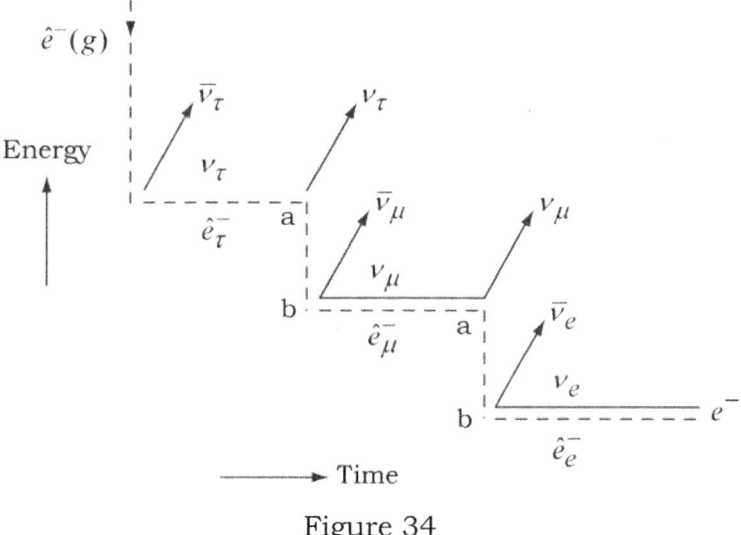

Figure 34

The leptonic pathway of decays and transformations

After tau decay, the intrinsic electron proceeds to a muon stage and finally arrives to a last ultimate stage – the electron ground stage of permanent stability, producing along the leptonic pathway two neutrinos (v_{τ^-}, v_μ) and three antineutrinos $(\bar{v}_\tau, \bar{v}_\mu, \bar{v}_e)$, as follows:

$$\hat{e}^-(g) \rightarrow \bar{v}_\tau + \tau^- ;$$

$$\tau^- \rightarrow v_\tau + \mu^- + \bar{v}_\mu ;$$

$$\mu^- \rightarrow v_\mu + e^- + \bar{v}_e .$$

What in SM is called a "decay", in AQM it is called something more constructive – it is the process of the transformation of one inner structure into another one, or even into several other inner structures. The "decay" is not destructive but constructive process toward energetically more favorable quantum state.

At the end of the leptonic pathway, we have the same original intrinsic electron $\hat{e}^-$ (g) that entered the chain of leptonic transformations beginning with tau stage and finishing with substantially reduced self-energy and unchanged helicity as a permanent electron constituent trapped inside the electron forever.

11.5 Leptonic Transitory Quantum Processes Between Stages of Relative Stability

Here is a detailed description of AQM spacetime dynamics of quantum processes taking place during two transitions between stages of relative stability: from tau to muon or from muon to electron shown in Figure 34 as (a-b)$_{\tau\mu}$ and (a-b)$_{\mu e}$.

A. Tau – muon transition

The following quantum processes take place:

1. Bound tau neutrino completes its drift toward the edge of intrinsic tau c-ring, crosses over the edge, thus unlocking corset force;

2. once over the edge, bound tau neutrino is freed from intrinsic tau repulsive force, additionally contracting by its inner self-corset force;

3. intrinsic tau expands radially with nearly the speed of light from tau quantum radius $r_{\tau q}$ ($r_{\tau q} < 10^{-22}$ m) to tau classical radius $r_{\tau c}$ ($r_{\tau c} = 2 \times 1.1 \times 10^{-16}$ m) within 3.7×10^{-25} sec;

4. intrinsic tau continues its radial expansion from tau classical radius $r_{\tau c}$ to muon classical radius $r_{\mu c}$ ($r_{\mu c} = 2 \times 1.86 \times 10^{-15}$ m) within 5.8×10^{-24} sec, releasing its energy, creating muon neutrino-antineutrino pair (ν_μ, $\bar{\nu}_\mu$), forming muon

$$\mu^- = \left\{ \begin{array}{c} \hat{\mu}^- \\ \nu_\mu \end{array} \right\}$$ at muon classical radius and releasing muon

antineutrino $\bar{\nu}_\mu$ which experiences additional contraction by its inner self-corset force; and

5. finally, following release of $\bar{\nu}_\mu$, the Corset action is triggered, reducing newly formed muon to quantum radius $r_{\mu q}$ ($r_{\mu q} <$ 10^{-22} m) and thus providing muon stage of relative stability with a lifetime of 2.2×10^{-6} sec.

The formation of muon is identical to the formation of electron. A detailed spacetime dynamics of electron formation is shown in Figure 27, Chapter 10.

The total tau-muon transitory time is equal $t_{\tau\mu} = 2\left(r_{\mu c} - r_{\tau c}\right)/c$.

B. Muon - electron transition

The transitory quantum process of muon – electron transition is similar to tau – muon process with a total muon – electron transitory time as

$$t_{\mu e} = 2\left(r_{ec} - r_{\mu c}\right)/c$$

173

11.6 The C-ring Length of Leptons as a Fundamental Constant

The subject of the c-ring length of the intrinsic electron is examined in Chapter 2.

Here we continue with our concept of the generalized intrinsic electron $\hat{e}^-$ (g) which could have any self-energy from zero to infinity. However, by now we know that g-intrinsic electron does not fall below the ground level of 0.511 MeV in the same way as the orbiting electron in the hydrogen atom cannot be found in orbit low -13.6 eV.

The g-intrinsic electron can be part of leptons or be a free traveling particle releasing self-energy and producing other inner structures, or be a particle with enormous self-energy of unknown origin.

I postulate here that the c-ring length of g-intrinsic electron l_c does not depend on its self-energy - it is self-energy independent.

Let us select any value of g-intrinsic electron self-energy. That would define c-ring, classical Compton radius, and classical Compton angular velocity. Then, by applying the General Compton Conditions and computing various values of the c-ring length, we can find a unique value of l_c by extrapolation. *The c-ring length l_c is a fundamental constant belonging to the class of fundamental constants such as c, e, h.*

The constant l_c is a unique parameter of the g-intrinsic electron. Another unique parameter is the g-intrinsic electron aphysical cylinder length L_a.

The mathematical procedure for obtaining l_c is described in Chapter 2.

Using computation based on classical electrodynamics, the c-ring geometry and the boundary conditions, one can compute electrostatic and magnetostatic field configurations, energy density distributions and total self-energy, and in the process confirming once again that electrostatic energy is equal to magnetostatic energy. There is only one unique value of l_c. A deviation from such value would require an adjustment to the of Planck constant $\hbar$, which would bring us in conflict with the Planck constant of the photon.

Values of neutrino self-mass, weak electric charge w, and weak electric Planck constant are established in AQM with surprising accuracy.

11.7 Intermediate Vector Boson $w^{\pm}$ Interpretations: SM versus AQM

A. SM interpretation

Intermediate $W^{\pm}$ vector boson plays a key role in leptonic decays and transformations. In particle physics textbooks, such leptonic interactions are presented in Feynman diagrams, shown in Figures 35 (a, b, c).

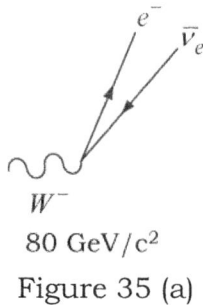

W^-
80 GeV/c²
Figure 35 (a)

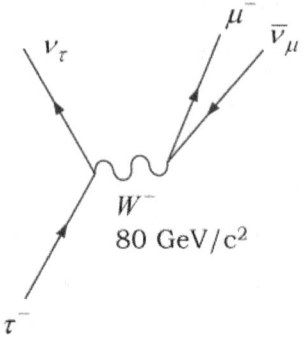

Figure 35 (b)

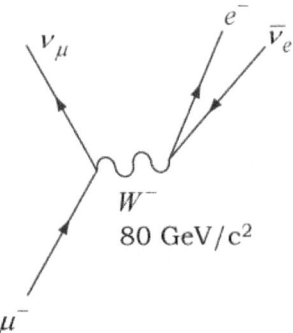

Figure 35 (c)

SM leptonic decays

According to SM, $W^{\pm}$ boson basic properties are:

Virtual energy: 80.355 GeV

Electric charge: ±1 e

Weak isospin: ±1

Half lifetime: 3×10^{-25} sec

W boson and Higgs boson are key ingredients of the Higgs theory and the theory of the electroweak unification.

B. AQM interpretation

W boson interpretation is another SM fundamental misconception.

Boson always travels in free space with the speed of light. The origin of W boson energy of 80.355 GeV has no explanation. Such energy cannot be borrowed from the vacuum for a short instant and then returned back to the vacuum to satisfy the uncertainty principle. I call such energy as virtual energy. However, virtual energy does not exist. It violates the energy conservation law. This whole concept is another SM fundamental misconception. Perhaps, what is called "virtual energy" is a value of energy at which the production of $W^{\pm}$ bosons is maximized in high energy proton-proton collision.

$W^{\pm}$ is not a boson. Bosons do not carry charge. "Charge" is the c-ring of the fermionic inner structure.

Then, what is W^{-} ? *It is the generalized intrinsic electron.*

♦ ♦ ♦

Postulate (theorem) 73. *The relationship between self-masses and between magnetic momenta of two lepton constituents, intrinsic lepton and lepton neutrino.*

At the lepton formation, classical Compton angular velocities of both lepton constituents, intrinsic lepton $\hat{l}$ and bound lepton neutrino v_l are equal:

$$\omega_c(\hat{l}) = \omega_c(v_l)$$

resulting in specific relationships between their self-masses and between their magnetic momenta as follows:

177

$$m(v_l) = \beta \; m(\hat{l})$$

$$M(v_e) = 2\beta \; M(\hat{l})$$

where $\beta = \frac{1}{2} \, \alpha$ and α is determined experimentally.

Proof of the theorem.

Part I. The relationship between self-mass of intrinsic lepton and self-mass of lepton neutrino

$$m(\hat{l}) = 2\frac{\hbar\omega_c}{c^2} \; (\hat{l})$$

$$m(v_l) = 2\hbar_w\omega_c \; (v_e)$$

$$\omega_c(\hat{l}) = \omega_c(v_l)$$

$$\hbar_w = \beta\hbar$$

Result:

$$m(v_l) = \beta \; m(\hat{l})$$

Part II. The relationship between magnetic moment of intrinsic lepton and magnetic moment of lepton neutrino.

$$M(\hat{l}) = \frac{e\hbar}{2m(l)} \qquad\qquad M(v_l) = \frac{w\hbar_w}{m(v_l)}$$

$$m(\hat{l}) = \beta \; m(v_l)$$

$$w = \beta e \qquad\qquad \hbar_w = \beta\hbar \qquad\qquad \beta = \frac{1}{2} \, \alpha$$

$$M(v_e) = 2\beta \; M(\hat{l})$$

◆ ◆ ◆

11.8 Long-standing Problem with Neutrino Mass is Solved in AQM with Great Precision

According to SM, electron neutrino mass is zero although nowadays few physicists believe that. It is another SM fundamental misconception.

There are enormous ongoing theoretical and experimental efforts including a recent *KATRIN* and *MARE* experiments toward resolving the issue of the electron neutrino mass, with marginal success so far.

AQM, a deeper quantum theory, provides the answer to the neutrino self-mass issue with a great precision and in the straightforward way.

This is one of the greatest AQM triumphs.

The final result:

$$m(v_l) = \frac{\beta}{1+\beta} \, m(l)$$

$$\beta = \tfrac{1}{2}\,\alpha$$

$\alpha = 0.001\ 159\ 652\ 188$

$\beta = 0.000\ 579\ 826\ 094$

A. Self-masses for leptons

$m(e^-) = 0.510\ 998\ 950\ \text{MeV}/\text{c}^2$

$m(\mu^-) = 105.698\ 389\ \text{MeV}/\text{c}^2$

$m(\tau^-) = 1.776\ 86\ \text{GeV/c}^2$

B. Self-masses for intrinsic leptons

$m(\hat{e}^-) = 0.510\ 702\ 761\ \text{MeV/c}^2$

$m(\hat{\mu}^-) = 105.637\ 138\ \text{MeV/c}^2$

$m(\hat{\tau}^-) = 1.775\ 73\ \text{GeV/c}^2$

C. Self-masses for lepton neutrinos

$m(\nu_e) = 296.188\ 828\ \text{eV/c}^2$

$m(\nu_\mu) = 61.251\ 169\ \text{KeV/c}^2$

$m(\nu_\tau) = 1.029\ 673\ \text{MeV/c}^2$

11.9   Lepton Properties and Data

The lepton is a composite particle consisting of the intrinsic lepton $l(n)$ and the lepton neutrino $v(n)$.

Here we consider properties of each component separately and then in combination.

We use the following notation:

n=1 (electron), n=2 (muon), n=3 (tau).

A. Intrinsic lepton

- Self-energy

$$\hat{E}_E(n) = 2\hbar\omega_c(n)$$

where $\omega_c(n) = c/r_c(n)$, and $r_c(n)$ is lepton classical radius.

- Self-mass

$$\hat{m}(n) = 2\hbar\omega_c(n)/c^2 = 2\hbar c/r_c(n)$$

- Spin

$$\hat{S} = \frac{1}{2}\hbar$$

Only magnetostatic self-mass of intrinsic lepton contributes to spin. Spin is the same for all leptons.

- Magnetic moment

$$\hat{M}(n) = \frac{e\hbar}{2\hat{m}(n)}$$

Only magnetostatic self-energy of intrinsic lepton contributes to magnetic moment.

B. Lepton neutrino

Recall the notation, $v(1) = v_{e^-}$, $v(2) = v_{\mu^-}$, $v(3) = v_{\tau^-}$.

- Self-energy

$$E_V(n) = 2\hbar\omega_c(n) \qquad\qquad E_V(n) = E_H^+(n) + E_H^-(n)$$

where $E_H^-(n)$ is magnetostatic self-energy inside the c-ring and $E_H^+(n)$ is the magnetostatic self-energy outside the c-ring. The border line definition between two magnetostatic fields is region where longitudinal projection is zero.

- Self-mass

$$m_V(n) = E_V(n)/c^2$$

- Spin

$$S_V = 0$$

181

- Magnetic moment

$$M_v(n) = \frac{w\hbar_w}{m_v(n)}$$

Only one half of neutrino self-mass contributes to neutrino magnetic moment, namely the inside magnetostatic self-mass m_v^- .

Neutrino magnetic moment has no anomaly.

C. Lepton

- Self-energy

$$E(n) = \hat{E}(n) + E_V(n) = \left(1 + \frac{\hbar_w}{\hbar}\right)$$

- Self-mass

$$m(n) = \frac{E(n)}{c^2}$$

- Spin

$$S = \hbar/2$$

Spin is n-independent. All leptons have the same spin.

- Magnetic moment

$$M(n) = \hat{M}(n) + M_V(n)$$

$$M(n) = \hat{M}(n)\left(1 + \alpha(n)\right)$$

$$\alpha(n) = 2\frac{w}{e} \times \frac{\hbar_w}{\hbar} \times \frac{\hat{m}(n)}{m_v(n)}$$

D. Transitory time

$$\Delta t_T = \frac{2r_c}{c} \text{ seconds}$$

Table 11-1.
Summary of lepton data

Lepton(n)	n=1 Electron	n=2 Muon	n=3 Tau
Self-mass	$0.511 \text{MeV}/c^2$	$106 \text{ MeV}/c^2$	$1.778 \text{ GeV}/c^2$
Self-mass contribution by neutrino	$296 \text{ eV}/c^2$	$61.25 \text{ KeV}/c^2$	$1.030 \text{ MeV}/c^2$
Classical radius r_c (pre corset), in meters	$2\times3.86 \times10^{-13}$	$2\times1.86 \times10^{-15}$	$2\times1.1\times10^{-16}$
Quantum radius r_q (after corset, estimate), in meters	$\leq10^{-22}$	$<10^{-22}$	$<10^{-22}$
Lifetime, in seconds	∞	2.2×10^{-6}	2.9×10^{-13}
Transitory time $2r_c/c$, in seconds	2.6×10^{-21}	1.24×10^{-23}	7.3×10^{-25}
Spin	$\hbar/2$	$\hbar/2$	$\hbar/2$
Electric charge	-e	-e	-e
Weak electric charge	-w + w = 0	-w + w = 0	-w + w = 0

♦ ♦ ♦

Postulate 74. *Fundamental role of the neutrino.*

The neutrino corset action demonstrates a fundamental role of the neutrino in particle physics as triggering, binding, and stabilizing force.

At the formation of the lepton the lepton neutrino corset action brings lepton classical Compton radius down to quantum radius by at least ten orders of magnitude.

The self-corset action of free lepton neutrino brings it down by many orders of magnitude, thus making the lepton neutrino stable with extremely small cross-section of interaction.

♦ ♦ ♦

♦ ♦ ♦

Postulate 75. *AQM complete definition of the lepton*

A family of leptons consists of three members: the electron (e^-), the muon (μ^-), and the tau (τ^-), at three different self-energy levels: 0.511 MeV (the ground level), 106 MeV, and 1.777 GeV, respectively. Beyond, high order leptons cannot achieve measurable lifetime. The lifetime is so short that it cannot be experimentally separated from other transitory quantum processes.

The lepton l is a composite fermion consisting of the intrinsic lepton $\hat{l}$ and the bound lepton neutrino v_l.

The intrinsic lepton inner structure consists of the physical energy c-ring and the aphysical energy cylinder.

The bound lepton neutrino inner structure consists of the physical energy c-ring and the aphysical energy cylinder.

The bound lepton neutrino has duo configuration consisting of the intrinsic neutrino and the intrinsic antineutrino with opposite helicity. The bound lepton is a duo-particle (Majorana particle).

During the lepton formation at pre-Corset stage, the lepton has the classical Compton radius and the classical Compton angular velocity. After the Corset action, the lepton radius is reduced to the quantum Compton radius by at least nine orders of magnitude.

During the Corset action the lepton does not emit or absorb energy. Its classical self-energy is preserved.

The lepton has three Position Parameters, PP1, PP2 and PP3. The Position Parameters are irreducible random quantum parameters.

No two individual leptons of the same type are identical – each has a set of unique Position Parameters.

In case of muon or tau, its lifetime is defined by a bound lepton neutrino drifting along the intrinsic lepton c-ring from the initial value of PP3 to the edge of the c-ring toward more energetically favored quantum state. Once over the edge, the bound lepton neutrino is freed from the intrinsic lepton. The free lepton neutrino then experiences its own additional self-corset action, thus reducing additionally in size.

Lepton spin is $S = \hbar/2$.

The lepton has negative electric charge, $-e$.

Lepton self-mass $m(l)$ is a sum of self-masses of both constituents, intrinsic lepton self-mass $m(\hat{l})$ and lepton neutrino self-mass $m(v_l)$ with relations:

$$m(\hat{l}) = \frac{m(l)}{1+\beta}, \qquad\qquad m(v_l) = \frac{\beta}{1+\beta} m(l),$$

where $\beta = \frac{1}{2}\alpha$.

Lepton magnetic moment is a sum of intrinsic electron magnetic moment $\hat{M}(n)$ and lepton neutrino magnetic moment $M_v(n)$ with the relation

$$M(n) = \hat{M}(n) + M_v(n), \text{ where}$$

$$M(n) = \hat{M}(n)\left(1+\alpha(n)\right)$$

According to the Theorem, value of α is the same for all leptons:

$$\alpha = \alpha_e = \alpha_\mu = \alpha_\tau$$

The lepton has four fundamental constants: l_c, l_c^v, L_a and L_a^v, corresponding to lengths of c-rings and aphysical cylinders, accordingly.

♦ ♦ ♦

11.10   A Hypothetical Lepton of 4th Generation

According to AQM, the lepton family has three members, but there is nothing special about number three. In fact, the number of lepton members is open ended.

My prediction is that by sharpening experimental technique, one can find a fourth member of lepton family with the following approximate range of parameters:

self-energy   6-10 GeV, and

lifetime 10^{-18} - 10^{-20} seconds

The question as to why Nature has selected such values of self-energy for leptons has not been resolved:

self-energy 0.511 MeV 106 MeV 1.777 GeV

lifetime ∞ 2.2×10^{-6} 2.9×10^{-13}

The answer has tremendous fundamental significance and requires a sustained scientific effort.

Let us recall the history of electron energy levels in the hydrogen atom, such as -3.6, -3.4, -1.5, -0.85 eV.

This set of numbers had made no sense until Niels Bohr introduced the quantization of the hydrogen atom.

Table 11-A

AQM Particle and symbol	SM Particle and symbol
I – Electromagnetic	
Photon, γ	Photon, γ
Intrinsic electron, $\hat{e}^-$	N/A
Electron, e^-	Electron, e^-
Duo-electron, e	N/A
II – Weak Electromagnetism	
Phontino, γ_n	N/A
Intrinsic electrino, $\hat{v}^-$	N/A
Electrino, v^-	N/A
Duo-electrino, v	Neutrino, v

This table can be extended for all AQM leptons.

Postulate 76. *Planck constant in the aftermath of the corset action.*

During corset action there is no energy emission or energy absorption. Therefore, energy conservation law dictates the following relationship:

$$\hbar_c \omega_c = \hbar_q \omega_q \qquad (*)$$

with ratio $R = \dfrac{\omega_q}{\omega_c} = \dfrac{\hbar_c}{\hbar_q} = \dfrac{r_c}{r_q}$,

where $\hbar_c$, ω_c *and* r_c *are classical values at pre-corset action, and* $\hbar_q$, ω_q *and* r_q *are quantum values in aftermath of the corset action.*

Dehmelt experimental data provides estimated value of the electron radius [4] .

The relation (*) explains how "the quantum world meets the classical world".

◆ ◆ ◆

11.11   The Complete Understanding of the Corset Dynamics with the Electron.

In his 1988 experiments with electrons in a Penning trap, Dehmelt [4] estimated a value for the electron radius with upper limit less than 10^{-20} cm. This is the discovery of monumental importance. After all, as it turned out the electron is not pointlike – *it has size.*

As a reasonable approximation, we assume ratio $R = r_q / r_c = 10^{-10}$.

As explained in Chapter 6, a Planck constant value is defined by length of the electron c-ring,

$$\hbar_q = \hbar_c \frac{l_c^q}{l_c}$$

where l_c^q is the quantum c-ring length after the completion of the corset.

At the electron formation, the corset dynamics brings both, electron classical Compton radius r_c and classical c-ring length l_c to quantum levels r_q and l_c^q. It is a translation of the classical electron from classical space to the quantum electron into quantum space dimensions reduced by more than ten orders of magnitude while energy conservation law remains intact (see Figure 36).

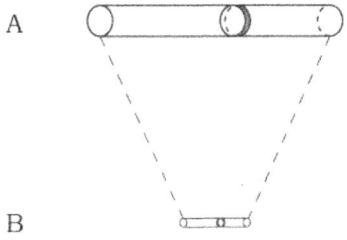

Figure 36

A – Classical electron c-ring

B – Quantum electron c-ring

◆ ◆ ◆

Postulate 77. *The complete corset dynamics during the formation of the electron.*

During corset process, the following electron properties are translated from the classical space to the quantum space, assuming $R=10^{-10}$:

Quantum c-ring length	$l_c^q = R l_c$
Quantum aphysical cylinder length	$L_a^q = R L_a$
Quantum Planck constant	$\hbar_q = R \hbar_c$
Quantum electric charge density	$d_q = \dfrac{d_c}{R}$
Quantum Compton angular velocity	$\omega_q = \omega_c / R$
Energy conservation	$\hbar_c \omega_c = \hbar_q \omega_q$.

◆ ◆ ◆

11.12 Conclusion

In 1947 Polykarp Kush and his collaborators performed experimentally accurate measurements of the electron magnetic moment and discovered a small deviation from Dirac's $g = 2$ by factor 1.00119. As it turned out, the discovery has had the disproportional impact on development and progress of QM, QED and particle physics. In the same 1947 year, Schwinger made calculations which agreed with Kush's measurements. This was the beginning of the bizarre scientific journey that has lasted to the present time and resulted in QED, the theory based on the dynamic vacuum teeming with activities and quantum fluctuations of virtual particles. For seven decades, the experiments and the calculations of QED contributions have been proceeding in parallel to ever-ever higher accuracy.

As of now, the experimental determination of the electron magnetic moment and the theoretical prediction agree to ten digit places. This is considered a triumph of QED, "jewel of physics - our proudest possession", as called by Feynman [1] (see page 176).

Not so fast!

In 1951 Dirac wrote,

"Recent work by Lamb, Schwinger and Feynman and others has been very successful ... but the resulting theory is an ugly and incomplete one", (see Ref. [1], page 176).

In 1987, Dehmelt's group performed experiments with individual electrons trapped in a Penning type device. Individual electron can be trapped for hours, days, even weeks, and subjected to extremely accurate measurements of its magnetic moment. This is an excellent example of AQM philosophy which is based on the description of individual quantum entities and individual interactions.

The experimental work has established the scientific credibility. It can be reproduced in other research laboratories with ever-refined experimental technique. The experimental work has proceeded along a single pathway with results converging to ever-accurate value.

This cannot be said about theoretical calculations, which are time consuming, requiring fast computers and many years of dedicated efforts (in fact, more than 20 years) by calculating QED contributions with ever higher orders of loop interactions.

The scientific credibility of the QED calculations (which I call "the theoretical experiment") has not been established for the following reasons:

- The underlying theory is bizarre. It is based on assumed existence of the dynamic vacuum filled with teeming activities and the quantum fluctuations of virtual particles. QED contributions are presented as an infinite chain of ever high end loop interactions where virtual photons perform acrobatics – they are emitted, then turned around and absorbed. It is obviously a statistical process. Although the experiments have shown that the anomaly is precisely the same for all individual electrons with the standard deviation zero to accuracy of ten digits. Somehow, the quantum uncertainty is not applicable to the dynamic vacuum filled with quantum fluctuations.

- The QED calculations have been running in parallel with the experiments. There has been no way to separate theorists performing the calculations from their awareness of experimental results. There is no way for us to evaluate influence of the experimental results on the calculations.

- A single and unique theoretical pathway has not been demonstrated. It would require a whole professional career life, in fact, several lifetimes, to accomplish this. I do not think one can find volunteers.

- Nature is presented as confused, chaotic, and cumbersome. Dirac was skeptical about QED contribution procedures. In his last paper, in 1984, on the subject of QED contributions to anomalous electron magnetic moment Dirac stated, "Just because the results happen to be in agreement with experiment does not prove that one's theory is correct", (see Ref. [1], page 176).

One can continue this analysis on and on, but it is not necessary. AQM has dramatically and fundamentally transcended QED and the Standard Model.

The QED calculations are based on the SM simplistic model of the electron. According to SM, the electron is the fundamental fermion of electromagnetism. In AQM terminology, "fundamental" and "intrinsic" are interchangeable. As AQM shows, Dirac theory of the electron is actually about the intrinsic electron, which has no anomaly in its magnetic moment. The anomaly is attached artificially to Dirac theory. In addition, the intrinsic electron is not stable.

The AQM electron, the real electron, is a composite fermion of electromagnetism and weak electromagnetism. The electron constituents are the intrinsic electron and the electron neutrino. The electron has the aphysical-physical inner structure of perfect geometry and a plethora of aphysical-physical properties, as shown throughout Volume Three.

The electron constituents, the intrinsic electron, and the electron neutrino are compatible – they have the same helicity, Compton radius and Compton angular velocity. The electron neutrino has a duo-configuration. It has a single field, magnetostatic field, spin zero ($\frac{1}{2}\hbar_w$ -$\frac{1}{2}\hbar_w$), weak electric charge zero *(w –w)*, and magnetic moment of two times of weak Bohr magneton. The electron neutrino performs its fundamental role at the formation of the electron by bringing the electron by its corset action from its classical Compton radius down to quantum Compton radius by at least ten orders of magnitude.

Both, the intrinsic electron magnetic moment $M(\hat{e}^-)$ and the electron neutrino magnetic moment $M(v_e)$ are aligned due to their identical helicities, producing the electron magnetic moment $M(e^-)$ as a sum of $M(\hat{e}^-)$ and $M(v_e)$,

$$M(e^-) = M(\hat{e}^-) + M(v_e),$$

where ratio

$$M(v_e) / M(\hat{e}^-) = \alpha, \qquad (*)$$

and α is the value of the anomaly measured by Dehmelt group (1987)[5].

The impact of such scientific discovery on particle physics and quantum physics in general is difficult to underestimate.

The relation (*) allows us to calculate properties of the electron neutrino, such as weak Planck constant, weak electric charge, and to solve the long-sought electron neutrino self-mass. This is the real triumph of AQM.

As I demonstrated in my theorem (see Postulate 73), the value of α is the same for all leptons.

$$\alpha_e = \alpha_\mu = \alpha_\tau = \alpha$$

Any deviation of α_μ or α_τ from α_e should be considered a systematic experimental error (faithfully reproduced by QED calculations as in the case of α_μ).

Foundational quantum theory, including the Standard Model, has been plagued for decades by quantum dogmatism mindset, as follows:

- *Quantum mindset I* is quantum positivism originated by the Copenhagen Group, especially by Niels Bohr and Werner Heisenberg. Since the Fifth Solvey conference in 1927, quantum theorists have never been able to get rid of quantum positivism. It has been detrimental to progress of fundamental quantum physics. Quantum positivists tell us that elementary particle has no complete definition. It is in all possible quantum states until it is measured.

AQM rejects quantum positivism. As an example, AQM tells us that self-entangled elementary particle has a defined quantum state configuration in spacetime. After measurement, particle acquires full-fledged quantum state, defined and objective.

- *Quantum mindset II* is the chaotic quantum reality, or CQR, originated around the year 1947 by a group of quantum pragmatists, including Schwinger, Feynman, Dyson, and Tomonaga in Japan, (see Ref. [1], page 175). The CQR mindset has "successfully" sidetracked the progress of quantum theory from its historical scientific trajectory (see Figure 37).

Here are some contents of quantum mindset II:

Quantum fluctuations, the indistinguishability of electrons (or any other elementary particles), quantum fields, virtual particles, virtual energy, the dynamic vacuum teeming with activities, the vacuum polarization, spontaneously emerging and disappearing particle-antiparticle pairs in the vacuum without cause whatsoever in violation of energy conservation law in each individual event, the universal wave function, the Hawking radiation, and the universality of quantum reality.

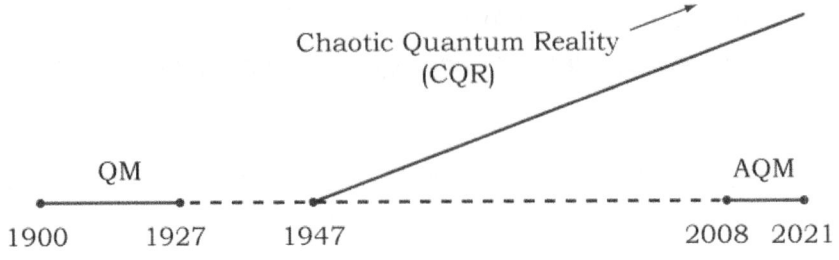

Figure 37

CQR – Chaotic Quantum Reality

1900 –1927: The First Quantum Revolution;
1947:   Kush measurements and Schwinger cal-
             culations;
2008 – 2010: The Second Quantum Revolution.

To support Chaotic Quantum Reality (CQR) requires dramatic expansion of role of mathematical formalism with associated extravagancy, virtuosity, and mathematical fog. Generations of theoretical physicists have been unable to be productive as a result of unmanageable complexity of mathematics.

AQM rejects CQR. As demonstrated in Volume One, Two, and Three, I succeeded to arrive at 27 fundamental scientific discoveries in foundational quantum physics and uncover 53 fundamental misconceptions in the Standard Model of particle physics.

According to AQM, Nature is elegant and majestic in its simplicity and sophisticated in its complexity. The intrinsic vacuum is eminently empty. Energy density in the vacuum is absolute zero.

This is what I mean by the Second Quantum Revolution.

PS: If the quantum pragmatists insist that they are able to calculate the electron neutrino magnetic moment, then, what stops them from calculating the intrinsic electron magnetic moment? Where are they going to stop?

If the quantum pragmatists insist that the quantum reality is the equivalent to the objective reality, then in three hundred years they would be able to calculate "the human", considered a subset of quantum reality, using computers, which are expected to be trillions times faster.

CQR is a myth!

References

[1]   Frank Wilczek, Fantastic Realities; 49 Mind Journeys and a Trip to Stockholm, World Scientific, New Jersey, 2006.

[2]   David L. Bergman and J. Paul Wesley, Spinning Charged Ring Model of Electron Yielding Anomalous Magnetic Moment, Gallilean Electrodynamics, Vol.1, 63-67 (Sept./Oct. 1990.

[3]   J.S. Bell, Speakable and unspeakable in quantum mechanics, Cambridge University Press, Cambridge, 1987.

[4]   [Dehmelt, H. (1988)]. "A single atomic particle forever floating at rest in free space: a value for electron radius is estimated". Physica Scripta. T22. 102-10.

[5]   R.S.Van Dyck, Jr., R.B. Schwinberg and H.G. Dehmelt, "New High-Precision Comparison of Electron and Positron g-factors", Phys.Rev.Lett, 59, 26 (1987).

[6]   D. Hanneke, S. Fogwell, and G. Gabrielse, "New Measurement of the Electron Magnetic Moment and the Fine Structure Constant", Phys.Rev. A 83, 052122 (2011).

[7]   G. Gabrielse, D. Hanneke, T. Kinoshita, M. Nio, and B. Odom, Phys. Rev. Lett. 97, 030802 (2006); 99, 029902(E) (2007)

[8]   Tatsumi Aoyama, Masashi Hayakawa, Toichiro Kinoshita, and
        Makiko Nio (2012). "Tenth-Order QED Contribution to the Elec-
        tron g-2 and an Improved Value of the Fine Structure Constant",
        Physical Review Letters. 109 (11): 111807.

[9]   Patrignani C.; Agashe, K. (2016). "Review of Particle Physics"
        (http://inspirehep.net/record/1489868/files/openspaccess cpc
        40 10 100001.pdf) (PDF).

[10]   T. Aoyama and 104 other co-authors from 82 institutions, The
        Anomalous magnetic moment of the muon in the Standard
        Model, arXiv:2006.04822v1 [hep-ph] 8 Jun 2020.

[11]   Victor Vaguine, The Second Quantum Revolution. Volume 1.
        Foundational Transformation of Quantum Mechanics. Aphysical
        Quantum Mechanics as Deeper Quantum Theory. Elementary
        Consciousness of Elementary Particles. ConsReality Press, 2020.

[12]   Victor Vaguine, The Second Quantum Revolution. Volume 2.
        Foundational Transformation of Quantum Optics. AQM Theory
        of the Photon. ConsReality Press, 2020.

About the author

Victor Vaguine, Ph. D., is the sole founder of Aphysical Quantum Mechanics, a deeper quantum theory and the origin of the Second Quantum Revolution.

Victor Vaguine is an independent scientist and philosopher.

He was born and educated in Russia (the former USSR). He earned his master's degree in Radiophysics (Electromagnetism) from the Gorki State University, Russia (1959).

From 1966 to 1971, he participated in the joint CERN-USSR program for development of equipment for high energy physics experiments at the European Center for Nuclear Research (CERN) in Geneva, Switzerland.

He received his Ph.D. in Physics from the University of Paris, Orsay, France (1970).

In 1971, after arriving to the United States, he accepted a scientist position at Varian Corporation, Palo Alto, California, an established high technology company specializing in radiation therapy equipment. A few years later, he was promoted to manager of a research & engineering division. He later left his secure position at Varian to take an active role in startup medical instrumentation companies.

As inventor and co-inventor, Vaguine holds 22 patents (United States and international). He has published numerous scientific papers and collaborated with several universities.

Since 2008 he has focused his scientific efforts on foundational issues of quantum mechanics and has developed Aphysical Quantum Mechanics, a deeper quantum theory, thus fulfilling Einstein's dream.

Aphysical Quantum Mechanics is the origin of the Second Quantum Revolution with immediate foundational transformation of quantum mechanics, quantum optics, and elementary particle physics. He has published his scientific work in three separate volumes under the title "The Second Quantum Revolution".

Victor Vaguine lives with his wife in Dallas, Texas USA.

His professional information can be found on *vaguine.com*

His email address is *aphysical.qm@gmail.com*.

Index

C

D

E

F

207

P

Q

R

S

T

V

W

Z

Made in the USA
Monee, IL
07 July 2026

56544785R00116